Swimming in Tech Debt

Practical Techniques to Keep Your Team

from Drowning in Its Codebase

Lou Franco

ISBN: 979-8-9923114-1-9

Cover Design by Michael Trent
https://www.michaeltrentdesign.com/

Interior design by Booknook.biz

To my wife, Jennifer

Contents

How to Swim in Tech Debt

Working in a codebase with a lot of debt is like swimming upstream: it drags me back from my destination. I eventually get there, but everyone else just sees the result. They don't feel the resistance. If I get slowed down, it just looks like I'm a slow swimmer.

But when I swim in a pool, each lap starts with a push off the wall. I get a burst of speed that I can extend underwater. Then I come up for air and swim until I get to the opposite wall and get to push off again. The push overcomes the water's resistance and periodically generates momentum.

My goal is to swim most of the lap on top of the water with freestyle strokes. That's the workout. But just like when I'm programming, it's more fun to start with some speed.

That's what drives my definition of debt. I don't think of it like financial debt that I owe; I think of it as code that resists my progress. I'm more likely to call a problem *debt* when it's internal to the codebase. It's something you can only feel when you're inside it, like water.

The scope of problems that I would characterize as debt varies wildly. It could be a bad variable name, an undertested module, a deprecated library, or the wrong programming language or architecture. Its main effect, I propose, is to slow us down, like when we swim upstream. It can have external effects too, but if that's all it has, I'd consider it a regular bug or feature request, not debt.

The core thesis of this book is that you can use tech debt payments to gain momentum while you code. The cadence of pushes followed by a nearly full lap of coding helps me finish the task at hand while leaving a series of tech debt payments in my wake.

In the pages to come I will show you how to pay debt while you are coding, how to make sure your whole team swims in the same direction to fix bigger problems, and, if you're in charge, how to set that direction and monitor it so that your teams can make the right tech debt decisions on their own. To support my claims, I'll share the latest research on software development productivity and how my recommendations align with their findings.

For some, "swimming" in tech debt implies drowning in it, and if you do nothing, that's what it feels like. But swimming also implies forward movement, of overcoming strong forces of resistance. *That's* the connotation I want you to embrace. Over my long career this mindset has served me, and it still brings me joy when I embrace it.

Let me tell you how I got here.

My History with Tech Debt

I have a confession to make.

As a programmer, I have created a lot of tech debt. Worse, as a manager, I've often argued against projects to pay debt down. It feels good to admit that. But I have also paid down my share and directed others to do so. Doing that helped me get better at knowing when it will be worth it (and what "worth it" even means).

This book is a collection of the most useful practices to pay down tech debt I have adopted and recommend to others. They are all based on my personal experiences, so I think it would be helpful to know who I am, what my professional focus has been, and what my biases are.

I got my first coding job in 1992. I was working at Astrogamma, a company that made a DOS application that served over 90 percent of the foreign exchange options pricing market. Four years later, we were acquired by a private equity group while I was leading the engineering team. At that time, we started a port of our application to the web, and I got my first exposure to systemic and sudden technical debt. Unfortunately for us, our entire codebase (a single-user, desktop application written in C) wasn't suited for this.

In the late nineties I joined Spheresoft, where I built a system for creating rich web applications. I wrote the first line of code, so I was responsible for a lot of its future debt. We started in C++, but eventually we had to move our SDK to Java, which we did by wrapping the C++. Years later the company did a full rewrite in JavaScript using Node.js.

Eventually, in 2006, I landed at Atalasoft, the leading .NET Document Imaging SDK. I was the head of engineering when we were acquired by Kofax and needed to support their Java customers as well. Before the acquisition the technical choices we had made were fine for our corner of the market. Now we had more resources and an ambition to double our market, but our most important product, a web-based document viewer, was coupled to ASP.NET and needed a rewrite.

Finally, in 2014, I joined Trello as a principal engineer on its iOS App. I worked there during the company's acquisition by Atlassian. Our team more than doubled, and we tackled long-standing debt projects, like shifting from Objective-C to Swift.

In 2021, I quit Atlassian to become independent and work on my own projects. I also advise software companies on their engineering practices. Unsurprisingly, tech debt is a constant topic of discussion.

At each job I had, no matter how sensible the initial technical choices were, something would happen that would compel us to change our minds. Being solo hasn't changed things. I'm the only engineer working on my codebase, and I rewrote my React front-end using React Native when my business partner realized that the application needed to be mobile-first. That codebase was less than a year old. It wasn't messy or hard to work with, but it could not support the direction we wanted to go. We'd be swimming upstream forever if we didn't do something about it.

Despite my varied experiences, a few things stayed constant during my career: I worked exclusively on B2B software products, and whenever I was in a leadership position, it was either at a small startup or a developer tools company. This meant that even as a manager or head of engineering, I had to stay close to the code. A lot of my suggestions in this book are drawn from this experience. While I think my approach will work for any programmer, you might find my emphasis on automated testing, observability, and code evolution through refactoring and migrations do not apply if your corner of the industry is very different from mine.

Another constant is that I spent significant time at each company, working alternatively as an engineer, engineering manager, and technical executive, so I got to see how our codebases evolved. I had to live with my choices.

Spending at least six years with four different codebases taught me that technical debt was inevitable. Working in different roles taught me that the only way to manage tech debt was by having processes to deal with it at individual-, team-, and organization-level scales.

How This Book is Organized

This book has five parts.

The first is "Rethinking Tech Debt," and I recommend you read through it entirely. All of the other parts depend on understanding it. Here I lay out how I want you to think about technical debt and provide high-level guidelines for judging when to pay it off and when to live with it.

The next three parts are directed at how to approach tech debt as an individual contributor, as a team, or as an organization's tech leader, respectively. If you wanted to jump right to the part that seems the most relevant to your position, that's fine. But please don't skip Part 1.

Part 2, "Personal Practices," offers a collection of practices that individual engineers can use to pay off smaller amounts of tech debt. If you regularly commit code, this section is for you. My intention is to give you ideas for improving a codebase in ways that pay off almost immediately. Note that each chapter in this part is meant to be independent and stand on its own.

The third part, "Team Practices," is meant for engineering managers, tech leads, and teams who need to set up processes and plans for making systematic debt payments. Here I describe a process they can run on a regular basis to track and pay down tech debt. That said, the techniques covered in these chapters can also be used by individual engineers to plan bigger fixes. If you lead a team, I recommend reading the chapters in order, as each builds on the previous one.

The fourth part, "Leadership Practices," focuses on things the technical leadership of a large organization can do to support the efforts of their teams as they struggle to manage their debt. The goal here is to a) give leadership the tools to determine if their teams' tech debt issues are big enough to require executive support and b) offer ways they can help their teams and engineers doing the work. If you are a team lead in a large organization with several layers between you and the CTO, reading Part 4 will help you to learn how to influence them. The chapters in this part are also meant to be read in order.

The fifth and final part is meant to help you decide what to do next after reading the book. To do that, it groups the chapters into themes that attack specific problems. The lessons will make much more sense if you've already read the previous chapters, but when you revisit the book, it will be faster to come back to this final part first.

The appendices contain sample meeting agendas and debt scoring guides that support the team process outlined in Part 3. I also included links to useful online resources, like the spreadsheets I use to score technical debt. The agendas, scoring guides, and spreadsheets are based on material I use in my consulting practice.

Why I Wrote This Book

For the past twenty years I've been writing about software development on my site, in my first book (*Hello! iOS Development*), and for other publications. Tech debt and developer productivity have been frequent topics. When it became clearer and clearer to me that the best practices for paying tech debt more than paid for themselves through immediate productivity gains, I started to write the book. I thought Parts 1 and 2 would be the whole thing.

Turns out there is a lot more to say. As I showed my drafts to others over the past year I developed my ideas further, applying them to the experiences of team leads and executives. Parts 3 and 4 reflect my belief that tech debt is a problem that needs to be addressed at multiple levels.

But the main thing that drove me to complete this is that I truly believe paying tech debt can make you happy. To me, working in a codebase without resistance is fun. I feel fast— maybe faster than I really am—just like when I swim back to shore and catch a wave.

With all of this in mind, let's dive in.

PART ONE
RETHINKING TECH DEBT

There are a lot of ways that addressing tech debt might pay off in the long run. Tech debt slows down coding, knowledge transfer, code reviews, and testing, and it can be depressing to slog through. But I think you should first justify the work using its immediate benefits. The trick is to keep your focus on the value you are adding and to consider its effects on your ability to do that.

CHAPTER 1

Find Balance

I try to think about tech debt in a way that leads me to take productive action. With that in mind, I define "tech debt" as any problem in a codebase that works against a programmer when they are trying to implement the roadmap. If we don't need to change the code or if the change is easy to make, then whatever we may think of the code, it's not debt because the debt has no effect. This goes both ways. If the code is beautiful but our goals have changed radically, then this code is debt.

It doesn't matter how the codebase got this way; it only matters that it's affecting us *right now*. Its existence decreases our productivity, and so we can get some relief by paying it down now, provided we do it proportionately.

It took time to learn the importance of proportionality. As a programmer, I always wanted to attack any debt I came across. But as a manager, I had to make sure that our team delivered value to our stakeholders and, as a result, ignored debt more than I should have. I went back and forth between these roles over my career, so I got to make mistakes in both directions more than once.

I got a better instinct for how to deal with tech debt when I moved away from thinking of it like financial debt. There are a few important differences between the metaphor of tech debt and financial debt. Let's go over them.

For one, I don't think of tech debt as intentionally borrowing against the future. Some tech debt starts that way, but often it accrues simply because the world has changed. Even if your code incorporates your best ideas for how to solve a problem, your ideas will get better, and the problem will change. You can do everything "right" and still have bad code, so it doesn't help to judge the decisions that got you there. Learn from them, but it's counterproductive to dwell on them.

I also don't think of tech debt as a fixed obligation. Unlike real debt, tech debt doesn't require regular interest payments. No one is going to show up to repo our repo. Tech debt "interest" is paid by the developer only when they need to change the code. If the code works, we never have to pay any interest. And if the code isn't used, we might eventually delete it. In both cases, paying debt on it would have been wasted time.

Finally, financial debt is absolute, but tech debt definitely isn't: You either have a mortgage, credit card balance, or car payment or you do not. We understand that the amount matters—debts can be large or small—but no one would classify any of these legal obligations as "not debt." That's true even for "good" debt that you take on to buy an asset that outgrows the debt.

Tech debt, on the other hand, operates in a far grayer area. It's almost never clear-cut that it must be paid off. Often, in fact, different programmers on the same team may not even agree on whether some specific code is debt. There may be plenty of good arguments for ignoring it.

Personally, the kind of code I usually find myself excusing involves a system that is so convoluted everyone is afraid to touch it—but it works and just isn't that important. So there's no need to change it. Sure, we find a bug in it every now and then, and it's painful to fix it, but preemptively rewriting it just isn't worth the trouble. Even worse is when there are no original authors of the code around to help understand it. I just want to ignore this code, and in most cases I think I can. But this becomes much harder to do if my mindset is to believe any debt must always be paid off.

The takeaway is that if you think of tech debt like a loan you agreed to pay back, there's a good chance you'll pay more attention to it than you should. But finding a balance means you can't totally ignore it either, which I'll get into next.

Don't Do Nothing

In 2010, I was the head of development at Atalasoft, an imaging SDK company. I was obsessed with shipping and spent all of my time worrying about delivering the features in our roadmap. Over time, we got better at releasing on a regular schedule, which showed up in our revenue growth and eventually led to our acquisition. The downside, though, was that I gave almost no thought to tech debt. I would learn my lesson soon enough.

We were a small company with fewer than ten developers and in a competitive market. There were more opportunities than we could handle, but we were bootstrapped, so we had to stay profitable and couldn't just hire ahead of our revenue.

The stakes got even higher after we were acquired. We had two years to deliver an ambitious new product. As part of the process, we had to adapt our technology to extend our new owner's legacy desktop products to the web and mobile devices. There was a carrot: an earn-out bonus that was contingent on this delivery. They didn't have to tell us about the stick, which I think we all understood.

Delivering this new product would require us to focus on that alone, with little time to clean up old messes. This was my mindset back then: that it was either one or the other. I believed any time spent servicing technical debt would keep us from delivering on our roadmap. But this intense focus had unforeseen costs.

Those costs first came to my attention during an exit interview, when a departing engineer mentioned that our level of tech debt had contributed to their decision to leave. I initially thought it was just them, but during one-on-one meetings I started asking other team members how they felt about what their colleague had told me. The common response was frustration with me and my inaction when it came to doing anything about tech debt.

By then, I had been a developer for over fifteen years and had worked in codebases with a lot of debt, so I well understood how tech debt impeded progress and caused frustration. But now I was a manager. While I sometimes contributed code, I was still removed enough that I had forgotten what it was like to be thwarted by the codebase every day. To my team, I was part of the problem. They had been trying to tell me this, but I didn't get it. It took an exit interview with their colleague—who at this point had nothing to lose—to finally get through to me. Only then did I understand the depth of the issue and how our failure to pay off even the smallest amounts of debt was slowing everyone down.

I learned an important lesson: the cost of tech debt is borne daily by your team, and by ignoring it you risk damaging motivation and raising attrition. Even if you believe you have every reason to move forward without addressing tech debt, being an empathic manager requires you to at least do something, even if that's simply acknowledging the existence of the debt. Doing nothing—like I did—is not an option.

With my eyes now opened, we started making some changes. The most acute problems were with our build system and installer because they affected every developer and every product we delivered. Each product and its installer were built via a tangled bag of legacy code that needed constant maintenance. It wasn't a ton of code, so I approved a plan to rewrite it with modern tools. It was a small experiment, but it paid off right away in quicker Continuous Integration (CI) builds and an easier to modify codebase. Most importantly, I saw that it didn't derail our roadmap. Following that example, we took on other small debt-paying initiatives, all of them equally successful.

This taught me another lesson about paying off technical debt. Until then I had thought of it as something that *might* pay off in the long run. Might. This belief made it hard for me to justify doing anything about it when I had to deliver on short-term goals.

In the above case, though, our tech debt payments left us with a build that had faster feedback loops, required less cognitive load to fix, and didn't make developers frustrated when they had to add to it, which we did regularly. Additionally, these updates could be made with less code, so they were less likely to break something else. I had been wrong about having to wait for a benefit: paying off technical debt had paid off in productivity right away.

Don't Do Too Much

As eye-opening as my experience was at Atalasoft, I had plenty more to learn. I got my next lesson at Trello, where I was a principal engineer on the company's iOS team. The codebase was three years old when I joined in 2014 and had some understandable tech debt because of how fast the company needed to move. Over the same short period, Trello had gone from zero to six million signups, and all while trying to ensure a tight product-market fit. Our biggest tech debt issue was that we were built on frameworks that made it fast to build a simple app but held us back as our app got more complex.

We knew we had to address this, but any available time we had outside feature development was eaten up by our need to respond to Apple's constant updates to iOS. In 2013, the iOS 7 update completely changed the iOS design language and its network APIs. The next year, right before I joined, iOS 8 introduced presentation controllers that gave developers a lot of control over the animation when new views are shown. That was great in theory, but it also broke our navigation code and caused crashes. As these problems added up, our code started to feel antiquated.

In my role at Trello, I was coding again full time, so all of this stuff was in my face every day. Luckily, we were a tiny team, so my direct manager was also coding a lot. He understood all of our problems. Empathy was not lacking.

We did fix some of these problems, but we sometimes went too far. To deal with the presentation controller problem presented by iOS 8, we developed an entirely new paradigm for screen navigation inside the app *and* rewrote all navigation to use it. This approach was the exact opposite from what I did at Atalasoft, where I'd pretty much ignored all tech debt.

Unfortunately, this major rewrite turned out to be overkill. In hindsight we could have just lived with the code we had and fixed the crashes. Instead, we spent a few months designing and implementing a new, non-standard way of writing navigation code. We forgot a vital lesson that one of our cofounders, Joel Spolsky, wrote about in a 2000 blog post, "Things You Should Never Do, Part 1":

> We're programmers. Programmers are, in their hearts, architects, and the first thing they want to do when they get to a site is to bulldoze the place flat and build something grand. We're not excited by incremental renovation: tinkering, improving, planting flower beds.
>
> There's a subtle reason that programmers always want to throw away the code and start over. The reason is that they

think the old code is a mess. And here is the interesting observation: they are probably wrong. The reason that they think the old code is a mess is because of a cardinal, fundamental law of programming:

It's harder to read code than to write it.[1]

On the Trello engineering team, we were all very familiar with this article and quoted it to each other often, but we still made the mistake it warned against. The urge to rewrite a system is strong. We should have addressed the few complex navigation cases that crashed our code without doing a full rewrite.

My lesson? Make sure any tech debt payment is proportional to its value.

A Tech Debt Heuristic

As you can see, I've experienced both extremes when it comes to dealing with tech debt. As a manager, I was overly resistant to devoting time to addressing it. As an engineer, I was exposed every day to the problems high levels of debt can create and supported paying off more than necessary.

These two extremes get at a fundamental tension. There needs to be a balance in how you handle tech debt, but finding it often isn't easy.

To help me figure it out, I've come to rely on a heuristic. I ask myself: by reducing this specific tech debt, can I increase developer productivity to deliver business value right now? If I can't, I won't pay it down right away. This simple analysis has made a lot of tech debt work pay for itself.

---

1   https://www.joelonsoftware.com/2000/04/06/things-you-should-never-do-part-i/

Don't worry: I don't completely ignore tech debt that doesn't meet this standard, in this book or in my work life. We just need to take a more methodical approach to it.

Since so much tech debt does meet this standard, I will tackle it first, in the next few chapters and Part 2. In Part 3, I'll cover approaches to paying off debt that has longer-term effects.

CHAPTER 2

Increase Productivity
Right Now

I like to think of tech debt as resistance to the progress I am trying to make right now. I live near a beach, and tech debt reminds me of the undertow that stops me from getting back to shore while I'm out swimming. Just as with an undertow, you don't want to fight tech debt because that makes it worse. So, I don't fight against debt-laden code. If it's stopping me from making progress, I fix it.

One easy way I regularly pay down a little bit of tech debt is by making small cleanup commits as I work. I started doing this more intentionally after I read Kent Beck's *Extreme Programming Explained* in 1999, which introduced me to automated unit tests and continuous integration. Then, when I read Martin Fowler's *Refactoring*, I started to learn how to improve a codebase over time with very small, behavior-preserving changes checked by unit tests.

In both books (and many others in the genre, such as *Working Effectively with Legacy Code* by Michael Feathers, as well as Beck's recent follow-up, *Tidy First?*) the authors stress that accumulating technical debt is inevitable. They argue that the best way to curtail it is by constantly fixing it with small improvements enabled by unit tests and mechanical refactoring. I agree, but with the important caveat that you should do this only when you are already changing the code for some other reason.

Unit tests, refactoring, and continuous integration are ubiquitous in the kinds of software I write, B2B productivity applications. Making small improvements on an ongoing basis is common among my coworkers. It doesn't take long, and usually brings quick wins, like making the code more readable by describing how the code is supposed to work (via unit test).

When I was at Trello, the iOS team adopted Model-View-ViewModel (MVVM) so we could test view-logic. This gave us the immediate productivity benefit of being able to run view code repeatedly without needing to manipulate a running app through several screens to confirm that our changes worked.

Even in frontend web code, where automated testing is typically rarer, I currently use the React Testing Library to unit test.

These small improvements work well when the debt is small or made up of independent chunks.

The issue is what to do when technical debt gets bigger. I admit this is where I've struggled. My problem at Atalasoft was not that I was against making small improvements. It's that when my team was confronted with more significant debts, I would almost always defer paying it, prioritizing actions that I thought would bring benefits right now (like delivering features for the roadmap).

But eventually I realized something important: You *can* get productivity benefits right now even with bigger debt-paying initiatives. If you do things right, you will deliver current feature work faster and with higher quality. In fact, at this point I am suspicious of any technical debt proposal that can't deliver some developer productivity right away.

Tech Debt Payments that Deliver Productivity

Rewriting the build and installer at Atalasoft is a good example. It took a single developer about a month. When we decided to do the rewrite, we had a backlog of problems and new additions to that code that would have taken a lot longer. But when the rewrite was done a lot of the problems just went away, because the new system was easier to get right. Since it was unit testable, we could add to it and keep it working much more easily. So, the total time we spent on the installers was shorter, even with the rewrite. We got benefits later, yes, but what we got right away was enough to justify the payment. Too many tech debt projects, however, only promise benefits later.

A couple of years later, I experienced a similar situation at Trello. When I joined, we were just about to start our internationalization (i18n) project, which I took on for the iOS app. One of my tasks was to implement i18n-safe string interpolation and pluralization. We wanted to be able to take a string like "There are $CARD_COUNT$ cards in the list" and make it possible to (a) replace the number and (b) ensure the sentence was grammatically correct for any language. (This is harder than it seems given how many languages' grammar and pluralization rules are nothing like English.) Both of these i18n features were only somewhat supported in iOS, so I would need to implement most of the code myself. It was just string manipulation at its core, which would have been easy to test, but in 2014 our iOS app didn't have any unit tests or the infrastructure to run them.

Without that infrastructure, if I wanted to check that my code worked, I'd have to run the app and then tap-tap-tap until I got to a specific string, and I would have to do this for each kind of string I generated. But with unit tests, I could just list all the examples with their expected results and run tests in less than a second. Since we didn't have any unit tests in the project, I proposed to the team that I'd add them to our build and continuous integration. I wanted to make sure there wasn't some good reason we didn't have them.

No one was against unit tests, but they hadn't been a priority. Most of the code was UI or network code, which are harder to test. But the code I was writing was highly testable, so I added the unit test project to our workspace and wrote the string code. With the unit test project in place, the other developers added tests to their own work. For the six more years I was at the company, I saw the benefits of the tests compound over time, especially when it came to complex code like our sync engine. The long-term benefits were a bonus. But that's not why I added the unit tests; I just wanted to go faster immediately.

I started noticing whenever we made quick productivity gains. When we adopted a new app design system with a reduced set of fonts, colors, and other design attributes and specific rules for using them, the same process played out. Originally, each screen in the app used hardcoded values for its attributes, which would make for an inconsistent experience when they didn't match each other. We could have just updated those lines to new approved values, but it was the perfect time to make an abstraction for the design system itself. Doing this made it much easier to write UI code that matched the design, saving us time and frustration right away.

When I reflect on my entire career, starting all the way back to my first job, I can see clearly how paying tech debt made me more productive immediately, not just eventually. In that first job, for example, I was assigned the task of updating our UI framework to use less memory. The original had been trivial to write, and worked great for years, but it didn't work as our application and its UI got bigger.

This was my first job. The application was orders of magnitude larger than anything I had ever seen, and I had only done rudimentary source control in college. But doing this work made me faster at navigating the codebase because I had to go all over our codebase to make the change. It also trained me on our source control system and gave me a little intro to each window and dialog in the system. Back then, we released only once a year, so our customers had to wait for the memory improvements. But I became a much more productive engineer in this codebase immediately.

My experiences made it clear to me that paying tech debt could have unexpected benefits right now. But it also highlighted something else: that the benefits could compound even further if the fixes delivered value to other stakeholders.

CHAPTER 3

Couple Technical Debt Fixes with Value Delivery

It can be hard to talk about technical debt with people who don't work in engineering. The problems we see exist only inside the codebase, which is invisible to stakeholders. But it's the water we swim in.

Tech debt slows us down. It makes it risky to change anything. It feels unprofessional. We know that it's going to cause problems for us in the future. But those problems, even when they are significant, are internal ones.

Earlier in my career as an engineering leader, I would describe paying debt to stakeholders as an end in itself. In retrospect, I think this is a mistake. We take a risk when we present some newly paid off tech debt as a deliverable. Our colleagues outside of engineering can't do anything with this information—and truthfully, they may not care. To many of them, "refactoring" doesn't mean anything. If a marketer comes to your demo to take notes for a launch, you don't want them to leave with a blank sheet of paper because they didn't understand a word you said. To mitigate the risk of having your work misunderstood or undervalued, try to pay debt that enables changes that have a high priority to stakeholders, and talk about that—not the debt itself.

If you look closely, examples of how other departments translate their accomplishments are all around your company. Your CFO reports a financial statement with revenue, profit, and curated line items for context; they don't explain how they had to restructure their ledger-item encoding system. The sales department tells you about when they closed a deal, but pipeline discussions are internal. Technical debt is just like this—an internal process. Treat it as such when discussing it with non-engineering stakeholders and you'll be much better off.

You can also look at widely different industries to see the same phenomenon. Doctors tell you that your procedure was a success, not how they washed their hands. Politicians tell you about the pork they brought home, not how the sausage was made. Your local sports team gets a ticker-tape parade for winning the championship, not having a good practice. But, just like them, you must have good practices to win a meet.

Members of each of these groups talk about the nitty-gritty of process improvements, but only among themselves. We should do the same. Inside your engineering organization, celebrate your debt payments all you want. Share stories with the rest of your team to build a culture that values them. But our stakeholders and customers only want to know how your software is improving in a way that they can see.

This is crucial. All of the examples from my career I mentioned in the last two chapters involved delivering business value alongside developer improvements. The new build and installer I worked on at Atalasoft were much higher quality, from my perspective and from a user's; many bugs just went away with the simpler system in place. The unit tests I added at Trello, which made it possible to implement i18n string interpolation and pluralization, made translation easier and more accurate. But I didn't mention the tests in the demo, only the features I added. When we implemented the new design system, it impressed coworkers during a demo not because of all the cleaned-up hardcoded UI values, but because the design was more accurately implemented. Similarly, at my first job, my work updating our codebase led to reduced memory usage in the UI. That's what my work was judged on, not clever changes to the code.

QA understood us. Designers understood us. Product managers (PMs) understood us. Translators understood us. When we did this right, marketing and sales could come to our sprint demos and know how to turn our improvements into webinars and sales pitches. Executives could see alignment with their Objectives and Key Results (OKRs).

That's not to say that we should hide any technical debt. We need to talk about it among our peers and management. The bigger the debt, the more we need to do this. But, when you pay debt off, celebrate internally, and to others emphasize the value delivered to QA, designers, PMs, and our customers. This value can be seen, and thus appreciated, while debt is invisible to them.

There are times when you *know* that a particular debt should be addressed but paying it won't help deliver on any part of your roadmap. You think it's obvious that you should pay it, and the other engineers on your team agree with you. While you may think you could explain it to your non-engineering colleagues, it's not likely. But there are things you can do to make the impact of the debt more visible to them. This is an important strategy to keep in mind whenever we can't justify paying the debt with faster roadmap delivery.

CHAPTER 4

Consider the Total Time Saved

As I mentioned earlier, when it comes to deciding what tech debt to take on, my bias is to choose whatever seems most likely to bring developer productivity gains almost immediately—to help us progress the near-term roadmap. I hope that there will be long-term gains too, but in my experience, they are often over-emphasized. I worry about the St. Petersburg Paradox,[2] which states that we can always make an argument to invest in something if we add up future gains over an infinite future, no matter how improbable or small those gains might be.

That said, I also consider all the time working on a debt could potentially save in coding, knowledge transfer, testing, reviewing, deploying, and waiting (not in the future, but right now, as I'm working on the issue). I think it's uncontroversial to say that adding a unit test will make it less likely that QA will find a bug or that new code will add a regression. But I have also found that adding unit tests to my Pull Request (PR) makes reviewing it much faster. This isn't just because it's more likely that the code is right: it's also easier for the reviewer to know that it's right. There are lots of reasons tested code might cause productivity gains in the future, but a faster code review pays off right away.

The same is true for adding comments to old code. If it takes time for me to figure out the code I need to change, it's going to take time for the reviewer, too. So, I explain my learning by

---

2   https://en.wikipedia.org/wiki/St._Petersburg_paradox

placing new comments around the code I am changing, speeding up the review. Doing this also allows the reviewer to catch a mistake in my assumptions that might not be obvious in the code change, another benefit. I prefer this to putting comments in the pull request because those go away when the review is done.

No matter how short the review cycle is, it's probably still going to take a few hours. So, when this PR is one of a series I need to do for a feature, while I'm waiting, I like to start working on a new PR in the same area I'm working on. Doing this allows me to stay in a coding flow without context switching now and then again when the original code review is done. Removing that overhead feels like I am creating time that otherwise would have been wasted and is likely to speed up the next feature PR.

Until a few years ago, all I had was my intuition that these practices were helping me be more productive. But in the last few years, as interest in understanding and measuring developer productivity has increased, various findings have supported my belief that the debt you should pay is the kind that is affecting your productivity right now.

Developer Productivity Models

There are a few developer productivity models that have been gaining traction in the last few years. DORA,[3] for example, has four metrics related to deployment efficiency and quality: how soon a written line of code is deployed, how frequently they are deployed, how often those deployments introduce critical problems, and how fast you can fix them. SPACE[4] recognizes that there is more to software development than just those four

---

3   https://dora.dev
4   https://queue.acm.org/detail.cfm?id=3454124

metrics, adding three more dimensions (including, notably, developer satisfaction) to its model. At the forefront of this work is Dr. Nicole Forsgren, who was the lead author of *Accelerate: Building and Scaling High Performing Technology Organizations* and has published a lot of research on developer productivity.

She and her co-authors have realized that most of the data measured by these two models are lag metrics, meaning that they show you the *result* of low productivity and thus aren't causal, leading indicators. Their recent work with the DevEx developer productivity model addresses that.

The DevEx[5] model is different in that it tries to express what drives productivity. It does this by identifying and measuring three drivers: flow, feedback loops, and cognitive load.

Staying in a flow state lets a developer do more things, because the more continuous time you spend in your IDE with the code, the more efficient you get. If you break up your time into smaller chunks, it takes time to get efficient again because you have to reload your short-term memory. My example just above of what I do while waiting for PR reviews shows one of the ways I keep myself in the zone. In these situations, I am trying to make good use of the feedback loop time and be productive while I am waiting.

That's why I rely heavily on DevEx to recognize when I am boosting short-term productivity by paying tech debt. That extra PR provides instant feedback that you are being more productive, whereas the DORA or SPACE models take more time to generate metrics (which means you might not know what exact activity changed the metric).

Here's another example. I mentioned that a good reason to comment old code before you change it is to speed up a code

---

5   https://queue.acm.org/detail.cfm?id=3595878

review, which shortens a feedback loop. Commenting is also a good way to reduce cognitive load. After spending time to learn what a section of code does, writing about it helps me remember what I figured out. Clarifying the code is even better. This whole process helps me get into a flow state faster than if I just stared at the code and cried.

We all know what it feels like to stare at code and wonder what it does, scrolling up, scrolling down, and then command-tabbing over to Slack to procrastinate. By adding short comments as I learn the code, I get more confidence that I understand it, and soon I find myself going from adding comments to making more substantive changes. This isn't for some future benefit—the comments are helping me *right now* because they drive productivity.

Staying in flow is so important to my productivity that I act as soon my code distracts me. Whenever I feel resistance to a change I'm making, I take immediate action to swim through it rather than avoid it. This often means making a smaller (but still useful) change or doing something else that ensures my flow will not be interrupted. Adding a comment or a unit-test is usually enough to keep myself from getting distracted.

But it's just as important to consider how your payment will speed up the entire process of getting a line of code into production. When you reduce the cognitive load of your reviewer or detect a regression before QA, you help make sure that your work won't get kicked back to you and pull you back from your next task.

Optimizing (and de-Optimizing) Feedback Loops

Of the three components in DevEx, I've had the biggest gains from optimizing long feedback loops. Extending flow and reducing cognitive load have immediate effects that are constant, but small. You can always swim faster by improving your form. Shorter synchronous loops work this way too, but the bigger effect comes from optimizing asynchronous ones.

Optimizing short synchronous loops is important. If you can get immediate feedback on your work, like when incorrect code is underlined as you type, you will go faster. Before my time, programmers used to "write" code by punching holes into cardboard and would get back compiler errors the next day. It's easy to see how reducing that loop was a boon. But now, we've got most things working this quickly. Asynchronous loops are the bigger problem.

There are many asynchronous loops in software development processes. We deal with code reviews and testing all of the time. But there are also loops with our product manager, designers, customers, and each other (e.g., when we request comments on our specifications). When these loops are longer than we expect, the difference between *work* time and *calendar* time starts to diverge. It makes projects late even when we coded as fast as we could.

In her seminal work, *Thinking in Systems*, Donella H. Meadows warned us that "The system, to a large extent, causes its own behavior! An outside event may unleash that behavior, but the same outside event applied to a different system is likely to produce a different result." She went on to say: "system thinkers see the world as a collection of 'feedback processes.'"[6] I take this to mean that a big reason that the review was slow was because of the work I submitted. The problem isn't the reviewer. The feedback was there to tell me what the problem is. It's me.

Technical debt exacerbates this issue, especially in code review, which is exposed to the same internal causes. If we work around debt, we might go faster, but we'll slow down the review.

---

6   *Thinking in Systems* by Meadows, Donella H.

No one wants to read that code, so it sits for a few days. Taking an hour to clean it up doesn't seem so bad now.

I used to think there was no downside, but that's not exactly true. Meadows goes on to show examples where feedback loops being slow actually helps. We run into this when a reviewer just comments "Looks good to me" and barely reads the code. And then a bug in production makes us roll back. So, we need to be careful if we decide to measure it. But don't give up.

When I want to measure something and also reduce the incentive to game that measurement, I couple it with another one that would let me know if the first was gamed. To do this well, it should be hard to game them both.

One issue with short loops is that we won't have time to catch bugs. To counteract that, I would concentrate on DORA metrics like **change failure rate** (deployments that introduce a major bug) and **mean time to recover** from that failure. These kinds of problems are usually addressed with an incident response process that includes a post-mortem and root cause analysis. When you do this, check to see if you think the bug could have been caught before deployment, and then figure out what you could do to make that more likely in the future.

Here's an example. We sometimes released the Trello iOS app with a bug that spiked crash rates. We decided to take a look. One thing we saw was that these crashes did also happen in internal dogfooding, but not enough to get our attention. We calculated the number of times a new crash would have to happen to cause a rollback and saw that just a few in-house crashes would translate to big problem in release. This made us keep more code out of the release branch if it had not had enough time to get high usage internally.

The other way we intentionally lengthened feedback was to change from immediate to rolling deployments. Now, when we released an app, users would not all get upgrades at once. It would roll out slowly, giving us time to see a problem, pause the rollout, and fix it. Slowing down ultimately got the correct code out faster.

In the rest of the book, I'm going to suggest shortening loops most of the time. Remember that you can overdo it. Figure out how to monitor the problem it might cause and use that to find a balance.

CHAPTER 5

Aim for Substitution

When I decide that code needs to be rewritten to pay off debt, there is a big range of what that might mean. At one end of the scale is *refactoring*, which changes only the code and does not change the behavior of the software at all. At the other end of the scale is a radical rewrite that might not resemble the original at all. It might not even be in the same language.

If you're not careful, paying tech debt can introduce regressions, so I like to apply Dr. Barbara Liskov's substitution principle[7] to this work. The paper in which she originally introduced this idea was written about evolving a software system through Object-Oriented (OO) inheritance using subtypes, but I use the framework to think through any code evolution. The thought process behind the principle can be applied to any code change, whether it's a simple bug fix or a full-scale rewrite.

What is Liskov Substitution?

Liskov Substitution describes a safe way to replace an object with a subtype. According to it, a substitution can be said to be "safe" if, in addition to meeting any language syntax requirements, the subtype also matches its base type's *behavioral* properties. Those behavioral properties are colloquially known as preconditions, postconditions, invariants, and the preservation of immutability.

---

7   https://en.wikipedia.org/wiki/Liskov_substitution_principle

Preconditions are what we expect to be true before our code is run and typically refer to the arguments and program state. Any code, not just OO code, has this concept. For example, in a math library, we might require that the argument to a square root function be greater than or equal to zero. To preserve behavior, a substitute is allowed to remove preconditions, but it can't ask for more. So, it would be okay to implement a new version that could take negative numbers and return a complex number, but it would not be okay to stop supporting the square root of zero.

Postconditions are what we promise our function will do if the preconditions are met and refers to the return type, the resultant program state, and any side effects. **Invariants** are conditions that are always true for all functions (even future ones). Adding postconditions and invariants in a substitute preserves behavior but removing any does not.

The final behavior that must be met by a potential substitution, **the preservation of immutability**, means that if some part of your internal state is promised to never change, then new versions cannot allow mutability of that state. This allows client code to safely hold onto data without making copies.

To summarize: as long as you only remove preconditions, add postconditions or invariants, and preserve immutability, your new code is not likely to introduce regressions.

There are various ways you could make changes that follow this more (or less) faithfully. When I'm cleaning up a code mess, I follow substitution rules closely. Keeping them in mind lets me know the risks I am taking when I don't. Next, I'll go through the options you have.

Refactorings Preserve All Behavior

Refactorings are highly substitutable because they are defined to preserve all behavior. This definition, described in the book *Refactoring: Improving the Design of Existing Code* by Martin Fowler, is important, as he explained on his site:

> The term "refactoring" is often used when it's not appropriate. If somebody talks about a system being broken for a couple of days while they are refactoring, you can be pretty sure they are not refactoring. They are restructuring.
>
> I see refactoring as a very specific technique to do the more general activity of restructuring. Restructuring is any rearrangement of parts of a whole. It's a very general term that doesn't imply any particular way of doing the restructuring.
>
> Refactoring is a very specific technique, founded on using small behavior-preserving transformations (themselves called refactorings). If you are doing refactoring your system should not be broken for more than a few minutes at a time, and I don't see how you do it on something that doesn't have a well defined behavior.[8]

In this book, I'll often refer to "mechanical" refactorings to emphasize this point. In these cases, I mean things like renaming a function, or introducing a temporary variable, or extracting lines from a bigger function into a new one that you call. Something like rewriting a linear search into a binary one is not refactoring because it doesn't preserve all behavior. See Fowler's

---

8   https://martinfowler.com/bliki/RefactoringMalapropism.html

online catalog[9] for more examples, and *Refactoring* for step-by-step examples of how to apply them.[10]

I start with refactoring (even if I know it won't be enough) because it is easy and safe. But it's often not possible to stop there. Also, as I will discuss next, refactorings preserve behavior, but not syntax, so they are not appropriate for public APIs.

Internal vs. External Contracts

I have worked at two developer tools companies in my career. When you write code whose main users are programmers that will write code to call yours, you have to honor both behavioral substitution and also whatever code compatibility requirements your language has. For example, adding a new argument to a function can preserve behavior but breaks the builds. This is okay in your own code, because you complete the refactoring by adding the argument to the callers, but you can't do that for external users.

The requirements for responsible library backwards compatibility are beyond the scope of this book, but my personal beliefs are along the lines of what SQLite, Clojure, Linux, and htmx do[11], which is to never purposely break user code. This is the opposite stance of Semantic Versioning,[12] which explicitly allows APIs to introduce breaking changes.

Rich Hickey, the creator of Clojure, calls changes that don't break user code "accretion," which is very closely aligned with Liskov Substitution. There are several language features in Clo-

---

9   https://refactoring.com/catalog/

10 https://martinfowler.com/books/refactoring.html

11 https://sqlite.org/lts.html, https://unix.stackexchange.com/
    questions/235335, and https://htmx.org/essays/future/

12 https://semver.org

jure that support this (e.g., you don't need to worry about introducing mutability because Clojure datatypes are all immutable). Hickey's 2016 keynote address[13] for the Clojure/conj convention described accretion in detail and applies to any programming language.

In this book, I am mostly talking about code with only internal callers, which can all be updated when an API syntax breaks. Even the simplest refactorings, while preserving behavior, do not try to preserve function signatures (or even names). When dealing with public APIs, I recommend Hickeyws accretion techniques and the stricter definitions of substitution that include caller syntax compatibility.

Refactoring Is Not the Only Substitutable Mechanism

Fowler called non-behavior preserving reorganization "restructuring," but his definition of behavior preserving is much stricter than Liskov's. If I can't clear tech debt with pure refactorings, having a substitution mindset gives me guidelines for going further without introducing regressions.

This means that my new code will not preserve *all possible* behavior, just the ones in the contract. This is okay if the new behavior is obviously better than the old behavior. There are a few common ways to do this: by improving performance, by fixing bugs, or by rewriting code to match your team's current coding guidelines.

---

13 https://www.youtube.com/watch?v=oyLBGkS5ICk

Improving Performance Is Often Substitutable

Using a better sorting algorithm is not a refactoring, but it is almost always substitutable. If you have a contract to produce an output in a certain amount of time, then doing it faster is more than you promised, which is okay.

Because all other behavior will be preserved, I often start optimizations by taking the old implementation and moving it into my test code. I'll be able to add an enormous number of tests to confirm that the output is the same.

One thing to look out for is if a performance optimization has different tradeoffs than your current code. For example, a naïve Quicksort is slower than Insertion Sort if the data starts sorted. Another common tradeoff is to choose a speed optimization that uses more memory (e.g. memoization[14]). Whether this is okay or not depends on the promises you have made. Substitution-oriented changes shouldn't break promises.

Fixing a Bug Is (Almost) Always Substitutable

Sometimes fixing tech debt also incidentally fixes a class of bugs. This happened to me at Atalasoft when we rewrote our build scripts with modern tools (see Chapter 1). The new implementation made entire classes of bugs disappear. You could also imagine this happening in a C to Rust rewrite because memory lifetime bugs become impossible.

A bugfix necessarily changes behavior, but it is not a contract change, because you didn't promise to be buggy. Unfortunately, it is possible that users of your code have worked around or otherwise depended on a bug's behavior, which your fix now breaks. Some teams (usually library providers) need to wor-

---

14 https://en.wikipedia.org/wiki/Memoization

ry about this more than teams that have control of all of their code's callers.

Raymond Chen of Microsoft has written many articles[15] about why Microsoft often supports undocumented behavior (including bugs) in new releases. He explains that Microsoft has even checked to see which program is calling a function to determine if they should act buggy (or not).

But for most of us, fixing a bug is not going to cause problems, so I treat that as substitutable. This means that when I fix a bug, I would do so in a way that preserves all other behavior.

Expressing Contracts in Code

There are some languages that were designed to express contracts like invariants and pre- and postconditions directly (e.g.: Eiffel and Ada). I have personally not used these languages in production, but in most languages, you can get similar support with assertions. I do this, but not extensively.

My main way of encoding behavioral properties is through unit testing, which I think has a few advantages. First, while I use assertions to establish some obvious checks, I want to do a lot more, and coding the contract in unit tests makes the code easier to read. Second, I personally keep assertions live in production for many of my projects, so unit testing allows me to do things that are not possible in production. I use assertions as a last resort to protect against doing something worse.

Third, unit tests can encode way more checks than are possible with just argument and result checking, like having a large regression dataset. I have often used large datasets to ensure backwards compatibility. At Atalasoft, we had an enormous da-

---

15 https://devblogs.microsoft.com/oldnewthing/20031224-00/?p=41363
    for example

tabase of images in every format (jpegs, tiffs, pngs, etc.). In tests we used this dataset by calling every image editing function we offered with a range of values and generated an output set to compare against. This allowed us to make changes we wanted while knowing that we'd be alerted if we broke something. We did the same at Trello for our Markdown parser (storing the generated HTML). I am currently working on a stock-trading simulator for a client, and I have a similar dataset to make sure that past strategies aren't broken by new features.

Whatever way you prefer to do it, the important part is to represent these contracts in code and not just documents. There's more on this in Chapter 18 ("Turn Documents into Code").

Try to Do Large Rewrites as Successive Substitutions

I have rarely seen large rewrites work, and so I tend not to do them and often argue against them. But I did once participate in a nearly two-year project to rewrite a legacy system in a more modern way.

One important takeaway is that the new system was developed while keeping the old system working. They were running side-by-side, so we could verify that they had the same behavior along the way. Even before we cut over, we knew that each piece (and then the whole) would be a safe substitute.

I cover this in greater detail in Chapter 30, "Give Big Rewrites Enough Support (or Don't Do Them)," and the Pragmatic Engineer newsletter has a useful post about them, "Migrations Done Well."[16]

---

16 https://newsletter.pragmaticengineer.com/p/migrations

Substitutability Everywhere

Of all of the programming books and papers I have read, nothing has had a more profound effect on my work as learning about Liskov's substitution principle. I now think about substitutability for almost every line of code I write inside of deployed software. One of the biggest arguments against paying tech debt in working code is the chance of regressions, which is reduced by incorporating substitution-oriented practices. Adopting them will make it more likely that you'll get support to pay more debt.

CHAPTER 6

Pay Off Tech Debt to Make You Happy

No matter how much I talk about the costs and benefits of paying down tech debt, there's an even simpler, less quantifiable reason I do it: it gives me joy. I think this feeling is natural. Removing obstacles makes me feel like I'm in control.

So anytime I am trying to figure out if paying a debt is worth the cost in time and effort, I factor this in. My own well-being is important, and it would be a mistake for me to ignore it. The feeling of being able to control my environment motivates me and thus will make me faster.

(That said, this can be taken too far. I have sometimes found myself getting so caught up in a debt that's bothering me that I do much more work on it than makes sense. It's important not to overvalue these feelings.)

I've come to realize that the joy I get in paying a tech debt is directly proportional to the increase in productivity it yields. In the last chapter, I explained how the DevEx model measures productivity three ways: by increased flow, shorter feedback cycles, and lower cognitive load. When I can do any of those three things, it makes me happy.

I hate having to wait for a long feedback cycle during a slow build or a review that goes back and forth for a few days. Similarly, increasing my cognitive load by having to memorize a bunch of gotchas about my codebase while I work builds up my anxiety, because my brain is worried that I'll forget something.

I feel like Ellen Ullman put it perfectly when she described the frustration of debugging a problem and realizing: "And down under all those piles of stuff, the secret was written: we build our computers the way we build our cities—over time, without a plan, on top of ruins."[17] Those ruins have a way of interjecting themselves when you don't keep them in mind.

Long feedback loops and high cognitive load certainly lower my joy. But I am most frustrated when I can't get into flow. Even before DevEx recognized it, the benefit of flow on programming success was well studied. In the 1987 book *Peopleware: Productive Projects and Teams*, Tom DeMarco and Timothy Lister describe a multi-decade experiment they conducted to measure developer productivity and find things that correlate with it.

The popularization of the "10x programmer" likely came from this book, since the authors found the very best programmers to be ten times better than the very worst performers. What's often forgotten, though, is that the best programmers were only 2.5x better than the median one. Additionally, DeMarco and Lister found no correlation between productivity and education, years of experience, or programming language.

So, what did matter? Their office environment:

> The bald fact is that many companies provide developers with a workplace that is so crowded, noisy, and interruptive as to fill their days with frustration. That alone could explain reduced efficiency as well as a tendency for good people to migrate elsewhere.[18]

According to the authors, for the median programmer, moving to a job where they can work uninterrupted will double their

---

17 *Life in Code: A Personal History of Technology* by Ellen Ullman
18 *Peopleware* by Tom DeMarco and Timothy Lister

productivity. They write about phone calls and meetings, but technical debt causes interruptions, too. That giant function that I can't understand takes me out of the zone just as quickly as a phone call does.

In his influential book, *Flow: The Psychology of Optimal Experience*, psychologist Mihaly Csikszentmihalyi explains why flow can be such a powerful motivator:

> [W]e have all experienced times when, instead of being buffeted by anonymous forces, we do feel in control of our actions, masters of our own fate. On the rare occasions that it happens, we feel a sense of exhilaration, a deep sense of enjoyment that is long cherished and that becomes a landmark in memory for what life should be like.
>
> This is what we mean by optimal experience.
>
> [...] In the course of my studies [...] I developed a theory of optimal experience based on the concept of flow—the state in which people are so involved in an activity that nothing else seems to matter; the experience itself is so enjoyable that people will do it even at great cost, for the sheer sake of doing it.[19]

Point being, paying tech debt will keep you in flow, which will make you productive and happier. This has certainly been my experience.

But it's also important to keep Csikszentmihalyi's warning in mind. If an activity is "so enjoyable," he warns, "people will do it even at great cost"—meaning, they might sacrifice productivity. In other words, your attempt to be productive can become, well, unproductive. This has also been my experience.

---

19 *Flow* by Mihaly Csikszentmihalyi, p 17

My defense against this is to acknowledge the joy that also comes from paying down tech debt, but to timebox it so that I keep the work proportional to the task at hand.

When a debt project is big enough that I have to quantify its costs and benefits more methodically (as we'll discuss in Part 3), I try to quantify the agony my team is feeling about the code and compare it to the joy we'll feel when we fix it.

CHAPTER 7

Bring AI Into Your
Tech Debt Practice

I am writing this chapter in the Spring of 2025. I put it off as long as I could because I am sure it will be out of date in a month. Oh well. I still think it's worth discussing AI and the Large Language Model's (LLM) effect on tech debt right now. So here we go.

In my programming work today, I use Cursor and all of its AI facilities on each of my projects. My work is either in TypeScript (React/Node) for web-based applications or Python (mostly to use Pandas). These languages and frameworks are very popular in GitHub and StackOverflow, so the major LLMs have been trained on them. Consequently, the code they generate is very good but not perfect. When I talk to people who use less popular languages, however, that is not always true (at least as of early 2025). Your experience with AI may differ from mine because of this.

Will Vibe Coding Introduce Tech Debt?

I do not personally "vibe code" production software. Coined by Andrej Karpathy, this phrase refers to coding with AI in such a way that you "forget the code even exists."[20] In my systems, I read every generated line of code and accept or fix it. Vibe cod-

---

20 https://simonwillison.net/2025/Mar/19/vibe-coding/

ing has its place, and I am a proponent of using it for prototypes, specifications, spikes, etc. I would even use it as a starting point for production code. But I would only commit it to my repository after it has been code-reviewed and fixed.

When I talk about tech debt, I am often asked whether vibe coding will generate a ton more of it. There are two ways to think about this. If you are planning to never look at the code, I think you will eventually run into limitations in what the AI can do. For small projects, this may never be a problem. But if these limitations go away (which feels possible) and if AI can always make the changes you want, then by my definition of tech debt (as resistance to change or swimming upstream) it will no longer be "debt." The code may be hard for humans to maintain, but as long as it works and the AI can maintain it, you'll be debt-free.

This isn't possible yet. The code I have seen inside vibe-coded projects is complicated enough to confuse the AI when it tries to add to it. To be clear, I am not talking about bugs or security issues, which exist in human-coded projects too. I am just talking about the internal code quality.

The second way to think about vibe coding and tech debt—my preferred way—is that vibed code is just like any code written by someone else and that has to go through code review and editing. To me, it's a phase of code construction, and thus must be followed by a clean-up phase. In that way, it's not that different from the way I write code manually, which is first to get it working and then to improve it through successive refactoring and tests. The amount of tech debt the AI introduces into my project is up to me.

Can AI Pay Off All of Your Tech Debt?

A more optimistic take, which I have heard a few times, is that we don't need to worry about tech debt anymore because AI can clean it for us. In the sense that AI helps me do a lot of things faster, I agree that dealing with tech debt is easier with AI.

I have personally found AI to not be as good at simple refactoring as just using regular non-AI IDE capabilities. For example, renaming a function (and all of its callers) is trivial, fast, and perfect in VSCode, while using a prompt to do it is slow and often doesn't work. The same goes for extracting code out of a large function. I imagine that at some point the AI will just use the IDE tools instead of trying to regenerate the code (see the rise of the Model Context Protocol [MCP]), but until then I won't use it for mechanical changes.

But for generating unit tests, it's a different story. In my experience, using AI for this saves a lot of time. It's also good at suggesting tests I didn't think of. Like a lot of things, the "answer" is that in combination with other techniques, AI is a useful tool.

Additionally, AI can help us reduce the costs of either paying or not paying down a tech debt (how to determine which path to take is the focus of much of Part 3). If we don't pay the debt, AI can help us safely make small changes without breaking things. Alternatively, AI is good at paying off small pieces of debt by adding explanatory comments or running tests. So even if doing this is a wash in terms of time saved, I opt for making small payments instead of living with these debts. Often even small fixes increase my job satisfaction and short-term productivity, both worthwhile wins.

Is Code with Tech Debt Harder for AI to Modify?

Anecdotally, I have found that AI and I have similar limitations when it comes to maintaining bad code. But we are also both productive when the code supports our intentions. The two projects I'm currently working on have very low debt, which means that new features can often be implemented in a straightforward way. That's not just true for me; it's true for the AI as well.

For example, in the backend code of my web application, I use TypeScript decorators to define my database schema, GraphQL queries, and mutations. Every database entity looks the same. Every GQL function looks the same. With all of this in place I can usually prompt the AI to write a new entity with queries and mutations that perfectly match what I have already. It knows how to add in the right authorization checks, uniqueness qualifiers, delete rules, etc. based on its training, but it's using my code to make sure it does it the way I would.

But when I work on code with a lot of debt, the biggest issue is that there is so much history that I have to keep in my head. The high cognitive load is a big reason why the code is hard to modify. When code is missing context cues, it's hard to hand it off to a new engineer or expect an AI to glean that context.

A major source of this problem is bad names. I use my understanding of English words to understand variables names and so does AI. We both get confused by the same bad names. Unsurprisingly, good names help AI code assistants, because they were trained from natural language. When the names are consistent and match the domain, the assistants have a better chance of understanding what our program does and could do. If you find the names in your code confusing, a good first step before prompting is to fix that with your IDE's renaming commands.

In Part 2 of this book, I make the case that paying tech debt right before you are about to make other changes will make the entire task go faster than if you tried to work around it. Paying tech debt in this way is essentially free. Until recently I always made these payments by hand. But for more than a year, I have been making them with code assistants and had the exact same (positive) experience. AI helps, but as the code gets easier to modify, it helps even more.

How to Apply AI to the Techniques in this Book

In this book I have tried to stay away from step-by-step coding instructions and instead describe techniques that aren't tied to a language, framework, or IDE. So, in the chapters that follow, I won't mention AI specifically, but for every suggestion I make, I personally have done it before LLMs, and I continue to do it now assisted by LLMs. If you don't (or can't) use AI, every suggestion in this book will still apply.

To make sure this chapter stays up to date, I am maintaining an addendum on my website (https://loufranco.com/tech-debt-book/ai). I will also provide AI addendums to specific chapters where it makes sense to do so. You will be able to subscribe to changes if you wish.

PART TWO
PERSONAL PRACTICES

The first part of the book offered my definition of debt, my thoughts on when to pay it or not, and various ways to conceptualize debt. It was, for the most part, fairly general. In Part 2 I get a bit more specific, detailing various practices I use when I'm paying debt as an individual coder. My hope is that you find these practices as valuable and time saving as I have.

Every codebase I've worked on has had some tech debt. Sometimes I swam against the resistance and let it slow me down, but now, most of the time I try to do *something* to address it. To make sure I am not wasting time, however, I always keep in mind the three drivers identified by the DevEx productivity model. So, I can consider my productivity increased if I (a) lengthen my time in flow, (b) shorten a feedback loop, and/or (c) reduce my cognitive load. All of the practices outlined in this section achieve at least one of these goals.

CHAPTER 8

Start with Tech Debt

You know the feeling. You sit down at your computer, ready to work on a feature story that you think will be fun. You sort of know what to do, and you know the area of code you need to change. You're confident in your estimate that you can get it done today, and you're looking forward to doing it.

You bring up the file, start reading . . . and then your heart sinks. "I don't get how this works" or "this looks risky to change," you think. You worry that if you make the changes that you think will work, you'll break something else.

What you are feeling is resistance, which triggers you to procrastinate. You might do something semi-productive, like reading more code. Or you might ask for help (which is fine, but now you'll need to wait). Maybe you reflexively go check Slack or email. Or worse, you might be so frustrated that you seek out an even less productive distraction.

To change this behavior, you have to change your relationship with resistance. I've trained myself to do this. Now instead of defaulting to procrastination whenever resistance arises, I pay off a small local piece of tech debt. With practice, you can control your triggers. So, when you encounter resistance to getting started, make small commits to remove it. Here are some real-life examples.

Sometimes you can't even figure out what part of the codebase has the functions you need to change. You try searching based on your naming conventions, other code like it, or what it's called in UI, but nothing shows up. You try browsing in the folders where that kind of stuff is, but it's not there or it's not easy to see. When you finally find what you are looking for, take some small action that will make this search easier the next time, such as:

- Changing the name of the file
- Changing its location in the file structure
- Changing the name of the function
- Moving the function to where it belongs
- Adding a comment that uses your search terms

Other times you identify the function that needs to change but don't understand how it works. The variable names don't make sense, or there are no comments. Maybe the code is fine, but it's distracting because it doesn't follow your current conventions. Or maybe the function is hundreds of lines long with convoluted logic. You scroll up, you scroll down, but you still don't get it. In these situations, you can make the code more understandable by:

- Adding comments to undocumented code as you learn it
- Changing confusing names
- Reducing complexity in large functions by refactoring them into smaller functions with good names

And sometimes you're pretty sure what you need to do, but you're worried about breaking something in the process. When you check, there are no unit tests, so you waste time manually testing your changes over and over as you work. Consider instead trying to:

- Refactor the functions you are changing into smaller chunks

- Make the code more testable
- Add new tests

Obstacles slow you down, so by removing them you will gain more control of your environment. When you do your work and ignore the tech debt, however, it means your environment is controlling you, which is frustrating—and not always faster. Remember: the goal is to swim in tech debt, not drown in it.

Make the Changes that Your Team Culture Encourages and Accepts

Any tech debt you pay should be proportionate to the overall work you're doing. If you spend three days paying debt on a task that is meant to take only two hours, you are obviously overdoing it. Your team and manager might think it's a waste of time because that wasn't the work they agreed to during your planning, and now you might be risking the success of the overall goal.

Personally, I aim for about 10 percent of my commits to be small tech debt payments to existing code. I also plan to make at least one small payment commit at the beginning of any PR. This aligns with my team's values; you should align with yours.

I prefer these changes to be in their own commits because it makes them easier to review. Any subsequent commits that involve changing some behavior will also be easier to review because they will be in isolation. If I'm fixing a bug in a complex algorithm, I might need to change some variable names first to make the code clearer, so I don't want the bugfix code change to be mixed in a commit with that.

To make sure your changes don't slow down code reviews, you should move the code in uncontroversial ways, meaning ways that don't just look good to you. For example, adding a comment, adding a new test, or breaking out code from a giant function are probably always okay, but be careful about renaming things on a whim. Changing names deserves careful thought.

Naming is hard, but there are times when it's clear that a new name is better. One thing that happens to long-lived codebases is that the names of things in the code don't keep up with changes in the user interface or what the business calls things. Matching names in the code with the words that the rest of the business uses is generally a good idea. You'll get another clear win by making names more consistent with each other. (Of course, don't just make naming changes in public APIs that would cause breaking changes; if those names are wrong, that's something that needs to be addressed by your team or organization and is not an appropriate change for a random PR. See Chapter 5, "Aim for Substitution," for the discussion about accretion.)

Different teams will have different standards. That's why this is something that should be discussed, decided, and then documented in your style guide (see Chapter 21 for more on this topic). Whenever there is back and forth in a code review about this kind of thing, use it as a trigger to have a discussion at your next retrospective.

Paying Small Bits of Tech Debt is Easy to Do

When you begin a coding session, you might find that you aren't always in the mood. In his book *Tiny Habits: The Small Changes That Change Everything*, behavior scientist BJ Fogg explains why we often feel this way: "You need to have both motivation and ability [to do] a behavior, but motivation and ability can work together like teammates. If one is weak, the other needs to be strong. In other words: The amount you have of one affects the amount you need of the other."[21]

So, if you have low motivation to get started, the solution is to start with something easy—like adding a comment or a test. Once you get going, you'll become more motivated, then you can move on to harder things.

It admittedly took me a while to figure this out. But now that I understand it, if I feel any resistance to starting at all, I use it as a sign I should warm up with some easy, valuable commits first. Since my plan is to apply the flow that arises to implementing a new feature, I direct my warm-up work to the area that is going to change. The result is that I am now turning time that I would normally spend procrastinating into technical debt payments, which leads to flow and makes me more efficient. It's a virtuous cycle.

Along the way I am also turning learning time into commits that reduce cognitive load. I would have had to spend some time to learn the code anyway, but whatever I learned only ends up in my head for the duration of the task. Making some commits not only lets me share what I've learned with my team and future self, making code reviews and future work faster. Another virtuous cycle.

During this process I gain confidence in my understanding of the code and that I won't break it. My reviewer will also become more confident as my learning is passed onto them, and they can use this knowledge to verify my change. Or, if I'm wrong about something, they will be able to see my assumptions about the code and correct me. Either way will be faster than if I had kept that to myself.

---

21 Fogg Behavior Model: https://behaviormodel.org

CHAPTER 9

Make the Effects of Tech Debt More Visible

In 2019, I was working at Atlassian when the CTO, Sri Viswa-nath, declared the entire engineering organization was going to spend several weeks addressing company-wide reliability issues. He later explained that he felt we hadn't "allocate[d] enough time on our roadmaps to address technical debt which had accumulated [...] This included the need for reducing technical complexity, improving observability, and addressing root causes when problems occurred."[22]

In practice, this meant that all feature development at the company was suspended to make time for projects that improved reliability, including fixing technical debt. I decided to concentrate on the Trello iOS app's data synchronization system, which lets users make changes offline that are sent to our servers later. We didn't expect the sync system to be 100 percent reliable, but we had support cases that indicated things were worse than we thought. Since those cases were hard to reproduce, though, and we weren't sure how widespread the issue was, they weren't given high priority. Until now.

It would have taken me more than a couple of weeks to fully diagnose and address the reliability issues, so I decided to focus my time creating dashboards that would help us measure reliability. When I finished, we could finally see our overall error rate. Engineering and product management all agreed it was unacceptable.

---

22 https://www.atlassian.com/engineering/five-alarm-fire

We picked an error rate level that we felt was within bounds, and an even higher level that would indicate "excellence"; anything below the acceptable level was deemed "unacceptable." We treated an unacceptable rate like any other incident and, following Atlassian's playbook, wrote an incident response guide for it.[23] Over time, the most common problems with the synchronization system were fixed, and we reduced our error rate by 75 percent.

Before we had the dashboard we couldn't justify working on the sync system, because we didn't even know if there was a problem with it. After we built the dashboard, however, we could see exactly how many users were having each problem, so we knew the value we were delivering by fixing the system. More critically, we knew that we *had* delivered that value when our numbers improved.

Most teams have some kind of observability requirement, but they are usually geared towards detecting anomalies that are addressed immediately with small changes. The difference here is that I knew a large subsystem of our code had a lot of debt and I believed it was causing problems, but I didn't have proof.

Here's a simplified version of the dashboard showing an anomaly.

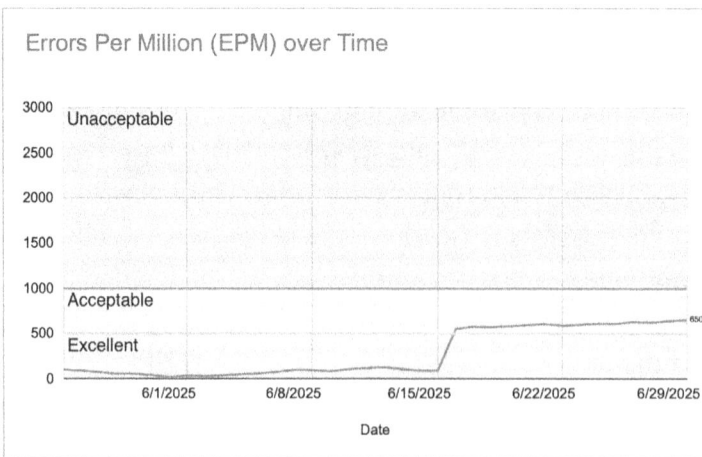

Errors Per Million (EPM) over Time

23 https://www.atlassian.com/incident-management/kpis/severity-levels

In it, EPM means "Errors per Million." We defined Excellent as an EPM below 500 (anything in the green area) and Acceptable as an EPM between 500 and 1000 (anything in the yellow area); a rating above the red line was Unacceptable.

We used the EPM metric because it makes it easier to see the difference between unacceptable and excellence than equivalent success rates. We wanted to go from a 99.8 percent success rate to above 99.95 percent. In raw EPM, that's 2,000 errors vs. 500 (a 75 percent reduction).

We also defined the response for each level:

- **Excellent**: No action required
- **Acceptable**: There must be at least one item in the current sprint with high priority to address this until the level is back to Green (Excellent). It can be deployed when the current sprint is deployed.
- **Unacceptable**: At least one person must be actively working to resolve the issue and doing hot fix deployments until the level is back to Yellow (Acceptable).

To their credit, our manager and product manager had agreed to let us fix things when reliability was Unacceptable, instead of risking items getting lost in the backlog.

Visualize the Leading Indicators of Productivity

If you can show the effects of tech debt on customers, do it. That is the most effective way to influence others. Most tech debt has only internal effects, though. In that case, I'd try to quantify the effect on developer productivity. Of course, to do this, you must be measuring it in some way. In my current advisory practice,

I've seen teams use DORA,[24] SPACE,[25] and DevEx[26] to gauge productivity metrics, which aren't perfect, but worthwhile. I recommend looking at all of them for ideas on what to measure. As I mentioned earlier, though, in my experience the best metric is one that reduces feedback loops because they are easier to quantify and control directly. I also think this is where the biggest gains are.

Feedback loops manifest themselves in a lot of ways, including the time it takes to build and test your system locally, how long it takes to get a PR reviewed, how long it takes for a CI run to finish, or how long it takes to get a design question answered. The shorter, the better. Few things frustrate me as much as a broken feedback loop. They are the main cause of gaps between my estimated work time on a project and actual calendar time spent on it.

If you are deciding on a metric, try to pick a leading indicator that you can act on right away in addition to lagging indicators that might only tell you months later that you improved. For example, a leading indicator like code coverage in a PR diff gives you feedback right away, but a lagging one, like the percentage of deployments needing a hotfix, while important to know, isn't something you can affect easily. But since the lagging indicator is often easier to associate with business value, this only works when the leading indicator builds up to the lagging one, so check to make sure they're in alignment.

Here's another way we made feedback loops more visible at Trello. A few years after I joined, our team had tripled in size. Our codebase grew significantly and so did our build times. After we complained that our laptops were too puny to do the work, we were issued Mac Minis with a lot more RAM. This

---

24 https://www.atlassian.com/devops/frameworks/dora-metrics
25 https://queue.acm.org/detail.cfm?id=3454124
26 https://getdx.com/research/devex-what-actually-drives-productivity/

helped, but not enough. But then, one of my colleagues wrote a small script that published our local and CI build times to our analytics database. To make them visible, he built a dashboard to show how much time we were collectively waiting for builds each day. The results were not surprising (to us), but now we had something we could show our bosses.

Since it could be argued that these delays were affecting every single developer on our team every day, it was easy for our manager to approve a new project to reduce build time. It turned out that a big part of the problem was the intersection of Swift and Objective-C code, so this gave us motivation to convert more code to Swift, paying off some debt. The effect of the conversion was visible on our build time dashboard, making the value of the effort understandable to everyone.

Add Tech Debt Detectors to Your Editor

Another way to make debt more visible is by analyzing the code itself. One way I do this is by using a combination of coverage and complexity. By themselves, they each only tell a part of the story, but together they give a good indication of the health of my code. Let's see how.

I can get code coverage when I run my tests. It tells me the percentage of lines that were run, and in the case where a line is a conditional, it reports how we cover the various permutations of that line.

Code complexity has many formal definitions, but for our purposes we can use the number of branches in a function. You should count conditionals and each case in a switch, but you should also count a loop as a branch. If you use a Boolean expression, you should count each Boolean operator as a branch. You don't need a perfect count, because we're just trying to categorize each function as "simple" or "complex," and once you get over the threshold you can stop.

The riskiest code to change has a combination of high complexity and low coverage, which makes it harder to understand and harder to verify. To make both of these metrics more evident to me, I use the Coverage Gutters[27] and CodeMetrics[28] extensions in VSCode. Whenever I am editing a function that is risky to change, VSCode shows a lot of red indicators around it. It looks like this:

```
Complexity is 11 You must be kidding
18   export const isAuth: MiddlewareFn<GqlContext> = ({ context }, next) => {
19     const authorization = context.req.headers["authorization"];
20     if (!authorization || typeof authorization !== "string") {
21       throw new Error("not authenticated");
22     }
23     try {
24       const parts = authorization.split(" ");
25       if (parts.length > 1) {
26         const token = parts[1];
27         const payload = verify(token, process.env.ACCESS_TOKEN_SECRET!);
28         context.payload = payload as any;
29       } else {
30         throw new Error("not authenticated");
31       }
32     } catch {
33       throw new Error("not authenticated");
34     }
35     return next();
36   };
```

The vertical dark red bars to the left of the line numbers show that this function isn't covered in testing. The red square at the end of line 18 shows that this function has high complexity (there's also a warning above the function).

27 https://marketplace.visualstudio.com/items?itemName=ryanluker.vscode-coverage-gutters

28 https://marketplace.visualstudio.com/items?itemName=kisstkondoros.vscode-codemetrics

This function is risky to change, but if I do need to change it, my first goal is to make one of those red indicators go to green. To do that, I can either add tests to get coverage or refactor to reduce complexity. If I can do it easily, I'll do both. In this case, it's easier to break the function down via a refactor than test it.

I can do this in a few steps. And then I end up with this function (see the green square at the end of line 18):

```
         Complexity is 5 Everything is cool!
18       export const isAuth: MiddlewareFn<GqlContext> = ({ context }, next) => { ▣
19         try {
20           context.payload = payload(authHeader(context));
21         } catch {
22           throw notAuthenticatedError();
23         }
24         return next();
25       };
```

It only took a few minutes. The new functions—payload(), authHeader(), and notAuthenticatedError()—are trivial to test, so I do that next. Now I have four simple functions mostly covered by tests.

I use these extensions to encourage me to make small fixes as I work, but the data could also be collected on the entire codebase to show areas that are risky to change. I recommend using this information in your time estimates, to make it clearer to stakeholders how tech debt will affect a project. When I triage issues that I am assigning to more junior engineers, I quickly check the area that is likely to change, and if it's undertested and complex, I point that out and ask them to fix that first.

To learn more about the usefulness of the combination of complexity and coverage, see this primer on it (https://testing.googleblog.com/2011/02/this-code-is-crap.html) from the Google Testing team.

Note the general technique of using two metrics in tandem to make it harder to game either one. If I test simple code to drive up coverage, I get no credit. I did something similar at the end of Chapter 4 ("Consider the Total Time Saved") with **lead time for change** and **deployment failure rate** (both from DORA). The first measures how fast a code change gets into production. The second tells me how often a deployment needs a hotfix. If you overdo the first one, the second will spike. Don't just pick speed metrics thinking you can go faster without consequences.

CHAPTER 10

Isolate Bugs into
Small, Tested Functions

Sometimes you know a function is doing something wrong, but you're not sure what the problem is. It may be a complex multi-threading or timing issue that's challenging to step through in the debugger. Or it just may be something beyond your current understanding or ability. When I work on an iOS app that crashes in production, often all I have to guide me is a crash log. I know the exact function that failed, but not why.

If you find yourself in this situation, consider using tech debt payments to guide your attempts to understand the bug. One option is to isolate the code in a small function that is obviously wrong and then fix it. To avoid creating additional problems, I do this via continual mechanical refactoring and automated tests.

If you go through this process and still can't find the bug, you can still squeeze out some external value if you add more logging in the area as you end the PR. If the problem you're dealing with is happening in the field but not on your machine, getting more information may be all you can do right now. But your stack traces and logs will be more valuable the next time because you've made the code simpler.

Example: Testing Sub-steps

I recently had to deal with a bug in a giant SQL query that pulled a lot of historical data to use in a visualization. I could see the problem in the bug report screenshot, but it wasn't feasible to use production data to reproduce it.

To diagnose the query, I broke it down into pieces that I could test individually. Then, I rebuilt the original query by composing the results of those functions. Next, I added tests to confirm that those smaller pieces resulted in the rows I expected them to return. Once the code was broken up, I could try more scenarios with each subquery without having to contrive tests through the original, much larger one.

I didn't find the problem right away, so I started generating more and more complex mock data. At some point, finally, one of my new tests failed because the sub-query didn't do what I expected. I had found the issue.

To finish the fix, I updated the tiny sub-query function. It was easy to see that this repair was obviously correct: the bigger SQL query based on it now worked *swimmingly*.

Make Sure Your Refactorings Are Mechanical

A word of caution: it's never a good idea to rewrite any code before you understand the issue you're dealing with. This is what I was getting at when I referred to "mechanical refactorings" earlier, which are different from rewriting or any kind of restructuring. I cover this topic in detail in Chapter 5, "Aim for Substitution." (Go back and read that if you need a quick refresher.)

A refactoring preserves behavior, including the bug. It's especially important to do that when we're diagnosing a bug because we don't yet know how to cause the correct behavior change.

I try to represent new knowledge as new commits. I could just step through the code and watch the variables manually, but if I do that via refactorings and new tests, I have something that (a) could be reproduced by everyone and (b) keeps checking the code forever. This preserves the knowledge for my team and my future self. I am also communicating (in the PR) that I understand the problem and making it clear that the final change (which fixes the bug or adds more logs) is correct.

That said, doing these refactorings could go on forever, so you have to get a feel for when to stop. If you think you are still learning about the bug and what may be causing it, then keep going. But if you are just moving code around because you like it better that way, it's time to stop.

When I'm doing this kind of work, I usually give myself a limited amount of time, with the goal of making a PR with better code and logging; I don't worry too much if I don't find the bug right then. Remember, this process is for fixing bugs that (a) have no way to reproduce on your dev machine but (b) occur often enough that you'll get back some logs quickly after the code is deployed. If so, getting some new code in the production application right away helps to shorten the feedback loop where we'll get more in-depth logs. Those better logs will save time for the next person that picks this problem up.

If we find the bug, all of the work we did to clean the code and test the area will make it less likely that this exact bug will resurface. And since all of our changes have been made near the old bug, it will get tested by QA, reducing the chance of us accidentally causing a regression by refactoring incorrectly.

CHAPTER 11

Fix Mistakes Twice

When I was a junior developer at my first job, Astrogamma, I was assigned a very strange bug. It was an easily reproduced problem, and the fix was obvious, but the cause was unusual. The code was in C, and someone had used a plus (+) instead of a logical or (||) operator in a Boolean expression. The code looked something like this:

```
if (booleanVar1 + booleanVar2 == TRUE)
```

This was before there was a native Boolean type in C, so like other teams, we used a char for bools and defined TRUE and FALSE to be 1 and 0, respectively. In simple cases, a plus actually works as an xor-operator, but the bug was showing up because we wanted to go into the conditional when both variables were true. Obviously, the person that wrote this had some misconceptions about Boolean expressions. We all make mistakes; I myself have written code so wrong I don't see how it could have ever worked. In modern shops, this would have been caught in review.

I remember that when I fixed this I wondered if the problem showed up in other places in the code as well. To find out, I made a bunch of elaborate regular expression searches; as it turned out, it was repeated in a couple of other locations. This experience taught me a valuable lesson, one that I apply every time I find a bug based on a programmer's misconceptions. I now assume that *any mistake I come across is only one exam-*

ple of a coding misunderstanding and, as a result, I immediately try to find and fix any others that the misunderstanding also caused. While technically this is just fixing a bug, not paying off technical debt, what I do next is.

If I come across a similar issue today, I turn my regular expressions into a lint rule. I didn't do this at Astrogamma, because linting wasn't as popular. If, instead, the problem is based on a misunderstanding of how a function worked, I would fix that directly by making the function signature clearer. Or if the problem is a third-party dependency that I can't change, I'd wrap calls to those functions in my own functions in a way that clearly communicate how the dependency works. All of these options make it harder to introduce bugs like the one I just fixed. Paying the debt makes it easier to write the correct code in the future.

If you are using a language with a powerful type system, it's likely that you can fix these kinds of problems using language features. In my opening example, just having a Boolean type that didn't define a plus-operator would be enough. Every time you find a bug that reflects a mistaken belief, you should make changes to the codebase that make that belief impossible.

The Localization Linter

When I was an engineer at Trello, one of my projects was to implement internationalization in our iOS app and manage the first round of translations that would use these features. To direct the translators, I needed to upload a file containing our strings and a short description of its context. The description and the string definition looked like this:

```
/* Notification that a card was moved */
move_card_notification = "$USER$ moved $CARDNAME$
to $LISTNAME$";
```

Translators would provide a new string, but they had to embed the variables (designated by dollar signs) in it. Sometimes a translator made a mistake, such as misspelling a variable or not including both dollar signs. Sometimes we made a mistake and didn't include a comment to explain the context.

"Fix it twice" was one of our mantras at Trello. With that in mind, I wrote a script to check for as many of these translation problems as possible. Every time we discovered a new translation issue, I tried to think of a way the linter could be improved to catch it early. Every error it caught saved a lengthy feedback cycle between QA, the engineers, and the translators.

Automating things was a good solution for us. But keep in mind that when you address a problem via automation, there is a startup cost to introducing a new tool and an ongoing maintenance cost to keep it working. So be sure the time you spend is proportional to its benefit. For example, I didn't make a new tool to find the wrong Boolean expressions earlier because it wasn't likely to happen again, but the incorrect usage of variables in a translation happened a lot.

Also remember that the practice of fixing mistakes twice has two parts. The first component, searching for other instances of a bug, makes you a super-efficient QA engineer, since you don't need to hunt for problems by driving the application. Often, a simple regular expression search in all files will be enough to find them.

The second component is the debt payment: trying to think of (and implementing) a process to make sure the problem can't happen again.

Whenever you try to fix a mistake twice, you might uncover more work than you can do right now. It's easy to go overboard paying this debt (and so many of the others described in this book) when we need to stay focused on the task at hand. Fixing the Boolean expression above probably took a few minutes. A quick regular expression search paid off quickly too. But if it had turned up hundreds of examples, I would have just submitted my simple bugfix and gotten through the rest in batches while waiting for code reviews. If the problem was enormous, I'd bring it to my team's attention so we could decide what to do together.

CHAPTER 12

Eliminate Debt to Make Estimates More Accurate

Estimating how long large projects will take is hard. Doing this before we've even gotten started can easily lead to an inaccurate estimate, but we have to try to determine if the project is worth doing at all.

Rather than just guessing, a better way to estimate a project is to build an extremely minimal prototype of it in a fixed amount of time. Doing this can help you move fast and explore the space you'll be working in, giving you a better handle on the project and come up with a more accurate estimate. By necessity, the prototype should take a lot of shortcuts, which allows you to try a lot of different approaches to solve the problem. As you go, ignore edge cases and most of the details. When the allotted time is up, stop exploring.

To guide this process, start with a simple breakdown of the project. This should include a high-level list of tasks, your best guess at how long each will take, and your level of confidence in each of those estimates. Use this as a map to help guide your exploration, concentrating on the areas where you have the least confidence and the highest time estimates, since that's where incorrect estimates will cause the biggest problems.

This process of exploring via a prototype is commonly called a project spike. The intention, again, is to throw away any code because you're just trying to develop a plan, identify risks, and make an estimate. Knowing that you'll throw the code away allows you to code in a haphazard way without the fear that you'll create more tech debt by merging the spike and hoping to fix it later.

You don't want to add your spike's debt to the product, but while you're working in your spike branch, you're likely to encounter existing debt that will affect the project you are trying to estimate. You could note it and increase your estimate, but don't stop there.

Just estimating how it will affect the project is helpful, but you'll be even better off trying to address the debt right now. To do this, make another branch where you just address this issue. If that change is obviously valuable, PR it now. If it might cause regressions, hold onto it until the project starts when you have more time to make sure it's right. You should do this even if you can only partially address the issue, because it will give you a better idea of the overall effect on your project.

Estimating big projects is a common source of anxiety. Sometimes it feels like just a guess or, at best, the summation of guesses. On its own this wouldn't necessarily be a problem; the issue is that you'll probably be held to whatever estimate you come up with.

One way to mitigate this is to have some real data, and that's the goal of the spike. Another benefit is that creating a prototype allows you to explore different approaches, frameworks, or algorithms without locking yourself into a solution yet. And then there's the opportunity to address some obviously wrong tech debt that we know will frustrate us later. Fixing this code now, during your exploration, means you'll be able to cut down on any delays later.

Finding Non-standard Code While Exploring

When I started working on the Trello iOS app, it had no centralized design system abstraction represented in the code. At some point, however, we standardized the design and built an internal library to implement it and help developers build compliant screens faster. As part of the process we did a pass to make sure the most important screens used it, but we didn't convert everything.

Several months later, I had to estimate a project that would affect one of these little-used screens, and I noticed that that screen's code had never been updated to use the design system's API. So I converted it during a spike because it was easy to do and check.

Since my goal with the spike was to create an accurate estimate, removing work that I would have to do later would serve that goal. That meant this work would be valuable even if my project wasn't ultimately approved. In the end my estimate was more accurate, debt was paid off, and I didn't use up any extra time (because spikes are timeboxed).

So I made a PR and wrote a ticket to make sure QA knew about it. I explained that the entire screen should be tested to verify that I didn't add a bug by mistake. There was no UI difference, as the issue just involved using hardcoded values instead of using the API. This allowed me to take a lot of "before" and "after" screenshots to include with the PR and issue.

Watch Out for Regressions

It's best to pay off technical debt that would otherwise prevent us from making changes to our product that are valuable to our stakeholders. This way, the debt helps us make progress toward a specific goal.

When we implement the feature after we paid the debt, and it's easier than it would have been, it lets us know that this time we've done it right. It also uncovers regressions (new bugs in working code) we might have introduced. If you build visible features on top of rewritten code, it makes it more likely regressions will be found by manual testing if the debt payments are in the same area of the application that changed.

Then again, a spike may not go anywhere. The risk is that we pay down a tech debt that makes it easier to make changes that we ultimately won't care about or ever need. Additionally, we may potentially introduce regressions that QA won't know to test. This would be worse than just wasting our time; it could potentially cause problems with customers if the regression causes bugs in features they need.

One way to mitigate that is to hold back the debt PRs until the project starts, so we have time to gain more confidence in the change. Perhaps paying the debt could be the first PR of the project.

If the debt payoff seems obviously valuable and uncontroversial (like my use of our design system API), then create tickets that make sure the change is tracked and tested before it is released. If you have manual testers, you should record a list of any areas that might be affected. Your descriptions should be written in a way that people outside engineering can understand.

Fixing Debt Now Will Lead to Longer Flow and Shorter Feedback Loops

When you encounter and fix technical debt during a spike, you won't be tempted to use up any more time because the spike is timeboxed. But if you're able to PR any code you generate during the spike, the project will be completed faster. You won't run into this specific problem while you're working on the project, so it won't distract you.

If you don't merge a fix during the spike, though, and instead wait until you encounter it during your project, it will either need its own PR (causing an extra PR review cycle) or complicate the PR you're trying to merge (causing a longer review). It's the same coding work, but dealing with the problem in isolation makes it faster to resolve. If all of this doesn't convince you to fix it, just remember to take it into account in your estimate.

CHAPTER 13

Make Progress While You Wait

I try to do most of my coding in pre-scheduled blocks of time that I've set aside in my calendar. I do this to maximize the chance I'll get into a flow state. Once I'm in the zone, I seem to go faster and faster carried by the flow. But sometimes while I'm riding a wave, I get to the shore before I wanted to.

Every now and then, I can't finish a feature because there's something I don't know, like a nuance of the specification that isn't clear. It's not something I can just research; it's something I need to ask about. I don't mind asking for help when I need it, but it means there will be a delay. For the past ten years I have worked remotely and asynchronously. During this time, the culture of my teams has been to not expect instant answers, so when I have a question, I usually need to wait a little while for a response. That's great, because that's the expectation I want others to have of me, but I'll have to tread water for a bit.

Other times, I know I can figure out something I have a question about, but I also know it's going to take a while. It could be a tricky algorithm, an API I'm learning, or a document that I need to reread. It has to be done, but in the moment the task might be a mismatch for my energy. Recall that flow is the feeling of moving without effort, and these things all require a lot of cognitive effort. In these situations, I find I'm better off trying to make *some* progress and then scheduling a block of time later to do reading, research, or heavy thinking.

Every once in a while, though, I have to stop coding because I finished up earlier than expected. Say, for example, I set aside three hours to finish up a pull request and complete it in one hour. I am still very much in the zone, and the time is blocked on my calendar, but I can't move on because I need the feedback from the PR review to make sure I'm going in the right direction. If I continue working without having received any feedback, I might end up writing code that isn't right and needs to be deleted.

My solution is to make a new branch and use this time, during which I would normally be waiting, to pay off some debt. In Chapter 8, I wrote about how I use tech debt payments to get into flow; it also works to extend it.

It's likely that I saw some debt while working. Once I am in the zone I concentrate on the work at hand, so I might ignore new debt I find if it's not too bad. But while I'm waiting around and still have a ton of energy for coding, it's the perfect time to tackle this debt.

One thing I started doing recently is to make very short-lived TODO comments. I personally try to keep the number of TODOs in my code at 0 (and I even use a lint rule to enforce this), but I am willing to use TODOs that don't get committed and are just for me. I think of every TODO as a kind of credit card debt that I need to pay within the grace period. If I want to ignore some debt I find while trying to get a feature done, I drop a TODO to address it right after I PR the new feature. I don't commit these comments; I migrate them to a new branch where I address them while waiting for the code review feedback.

Make Sure Your Context Switching Is Worth It

Context switching (such as what I just suggested above) costs time, so you need to be careful. The best use of your pre-allocated time and built-up flow is to direct it towards your larger goal, not some random piece of tech debt. So any debt you choose to tackle should ideally relate to what you are currently working on. If you make a mistake and cause a regression, you are more likely to uncover it if it's in the same parts of the program that have visibly changed.

As a counterpoint, many developers, including me, have adopted the practice of stacked diffs. This means that I make several PRs close together, each building on the one before it. All of these PRs (or diffs in this context) could be up for review at the same time, but the entire change in the whole stack is broken up to make each review chunk more manageable for your teammates.

The main purpose of this practice is to allow you to keep coding when you finish a PR. If your team culture supports stacked diffs and they work for you, it might be a better choice than going on a tangent to fix debt. But this only works if you can make the next diff in the stack without feedback. If you can't, then paying debt is a better choice.

At Atlassian we used BitBucket, which didn't have direct support for stacked diffs at the time, so it was sometimes technically difficult to deal with problems in the first PR that rippled onto the others. Our team valued small PRs, though, so I didn't want to just combine them; it was much better if the subsequent ones in a stack weren't highly dependent on the previous ones. As a result, I would often follow up a PR with new ones that paid off tech debt in the next area I was going to change. When the first PR was approved, I'd move on to my next change while the tech debt PR was getting approved.

Ultimately it may be worth going on a tech debt tangent when your last PR is likely to have comments that affect how you move forward, but the tangent should be deliberate. If you learn to adopt this practice, being forced to wait for comments will become a trigger for figuring out how to extend your flow.

As I outlined in Chapter 6, built-up flow is a source of joy, so extending it is a great way to build up satisfaction in your work.

CHAPTER 14

Remove Tech Debt When It's Your Decision

I'm a big proponent of incorporating tech debt payments into your everyday workflow. Ideally it feels like a natural extension of your tasks. To me, paying tech debt is part of delivering a feature, not something you do instead of delivering one.

Admittedly, this may make bigger tech debt issues harder to tackle, especially when they can't be broken up in smaller pieces. If the only way to fix something is to spend a big chunk of time on it, and the end result doesn't have a clear business value, it's not very likely that customers, product managers, or even engineering managers would think it's a good use of your time. You could try to build support for it by making the effects of debt more visible to these stakeholders (see Chapter 9), but that's not always possible—or successful.

There may be another option, however, if you have a strong feeling that a payment is warranted, but you can't justify it yet. Many companies have a ritual where they just let engineers do whatever they think is best for a set amount of time. At Atlassian, we had several regularly scheduled blocks of self-directed time. This included three ShipIt Days a year: fun, one-day hackathons where we could make whatever we liked. We also had occasional Innovation Weeks where we were encouraged to pursue and prototype ideas that we generated.

My personal strategy was to use this time to work on something I thought was valuable but a) wasn't simple and b) didn't yet have wide support. Sometimes I directed that energy towards bigger technical debt projects.

As I've mentioned throughout this book, the biggest lever you (as an engineer) have for getting buy-in is to reduce the cost and risk of a project, tech debt or otherwise. Often the reason we can't convince others to support a project is because they don't agree with our cost/benefit calculation. When you have a large block of unrestricted time to work with, it's more effective to use that to lower or confirm the cost side because that is within our control. It's not as easy to get others to believe that you have increased the benefits.

Use Hackathons and Other Special Events to Reduce the Cost of Paying Debt

When I started at Trello, a customer support colleague sat in on our bug triaging sessions. We did this so that she could help us understand the impact of each bug on our customers. At the end of each meeting, I would ask her what she thought the biggest problem in Trello was, not confined to the tickets at hand. She always said the same thing: people don't like that the Markdown in our cards rendered on the web doesn't match the Markdown rendered in our app.

Our app and the web rendered simple Markdown the same, but Markdown is poorly specified, so the library we used in iOS didn't match the web and could never match every edge case. Over time we had collected a lot of real-world examples that were totally reasonable and should have matched. Eventually I decided the only way to make the iOS Markdown parser perfect was to just port the web one, but I didn't have time.

When we got acquired by Atlassian, along with their management and systems, we inherited ShipIt Days. To me, this seemed like a great opportunity to pay off some of our Markdown debt. I was never ever going to be able to port the web parser to Swift in a day, but I wanted to see how far I could get during one of the sessions. Ultimately it helped me determine two important things: that my technique would work and how much longer the process would take to complete.

Looking back, I probably could have justified doing this work earlier if I had limited it to a day but being able to do something without any accountability at all unlocked something creative in me. During ShipIt Days no meetings could be scheduled, and if you were doing a project for the event people generally left you alone. Maybe not surprisingly, my focus and flow during this twenty-four-hour period was very high.

At the end of these hackathons, we were encouraged to present our findings to our team or the wider company. For me, this was always a great opportunity to bring attention to the benefits of paying down technical debt and, perhaps, secure more time for it because the costs had been lowered. If you have a similar opportunity, make sure to focus on the features that you will be able to implement faster when the debt is addressed. When I did this for the Markdown parser, I just showed the list of bugs and support cases that would be fixed by finishing the project. We scheduled the time to complete the project soon after.

Stay Within Bounds

When the Atlassian CTO stopped all feature work to give engineers time to fix reliability issues (as I discuss in Chapter 9), that's what we needed to work on. It just worked out that our reliability issues were in code that was hard to fix because of its debt. I could solve a problem that was on my mind and also address his concerns.

Another time, we had an Innovation Week that was clearly focused on exploring feature ideas, and so that's what I did. If I had instead worked on one of my pet debt projects, I would have made it harder to get leeway to do so in the future. In Part 4 of this book, I argue that C-level leaders should give autonomy to their engineering teams to pay the debt that they think is appropriate (under constraints.) One way to get this freedom is to show that you know how to use it when given a chance.

CHAPTER 15

Talk About Tech Debt During Job Interviews

Tech debt may be unavoidable, but you'll be in a better position to glide past the worst of it if you work with a team that values paying it down. The time to determine this is when you are interviewing for a job with them. I always try to figure out what the codebase is like during my interview. In Chapter 30, I'll tell a story about spending two years rewriting a codebase, but I knew exactly what I was getting into before I started because I asked a lot of questions.

At the end of any interview, you should expect to be asked if you have any questions. If tech debt is something you care about, this would be a good time to mention that. It's also a good time to discuss your approach to handling it.

Personally, I would frame my questions differently based on the job I was applying for and the person I was speaking to. In the sections below, I break down questions three different types of jobseekers might ask:

- Someone applying for a relatively junior position
- Someone applying for a more senior but still individual contributor position
- Someone applying for a team lead or engineering manager position

For each of those cases, you might be interviewing with a peer engineer or with the person who would be your manager. The way you talk about tech debt should be tailored based on your interviewer. Regardless, if tech debt matters to you, you should make that clear. (An exception: If you are talking to a recruiter or someone outside of engineering, you can probably guess from my earlier advice in this book I don't think you should talk about debt.)

Questions to Ask as a Junior Developer

If you are applying for your first job as a new grad, frankly it's surprising to me that you're reading this book. But I salute your interest, and I think it would be appreciated by hiring managers as well.

During your interviews, I would bring up the topic to show that you may have more experience that it seems, or to reiterate the part of your background that exposed you to bigger problems than you might encounter on smaller, learning-type projects. If you *have* worked on large codebases and with larger teams, talking about your experiences offers a great entrée into talking about tech debt.

You could say something like, "When I was working at my internship, it was a fairly big project, and I noticed *<insert codebase problem here>*. Is that something that happens here and, if so, what do you do about it?" The codebase problem depends on the project, but it could be something like, "there were no unit-tests," "the build time on CI was long," or "the dependencies were out of date and hard to update." You want to mention something that would be well understood by engineers, but perhaps not as easy to explain to others. Your goal is to discover their attitude towards dealing with these kinds of issues. Just as important, you want to communicate that you understand these issues exist.

Alternatively, you could signal your interest in tech debt issues by asking for their advice about an issue. For example: "While I have your ear, I'm having this tech debt problem that I'd love your input on . . ." By doing this, you are giving them an idea of what it's like to work with you. You show that you know that being a software engineer is more than what you do in the IDE.

A lot of engineers coming out of school have only worked on codebases that they wrote on their own or small school projects with a couple of others. Explaining that you've worked on bigger projects, especially ones that accrued tech debt that had to be paid off, could help you differentiate yourself.

Questions to Ask as a Senior Developer

As a more senior engineer, you will have worked on large problems and have had many encounters with tech debt, good or bad. If you believe your strategies for paying down tech debt have been novel or boosted productivity, you should talk about them.

As always, I suggest expressing the impact of your accomplishments in a way that speaks to your interviewer. If you are talking to a prospective peer engineer during an interview, I would just ask about the codebase directly. Specifically, what their team strategy is for dealing with tech debt. If you are talking to the CEO of a startup, though, you'd probably have more impact linking your debt payments to increases in business value. If you are talking to someone that values craft, lean more into that. Find common ground.

You can gauge their responses to see if their values align with yours. They will sense it too, so you should also be prepared not to be offered a job if there is a big gap. You know yourself, so if avoiding a job with mismatched values is a high enough priority, it's worth figuring that out. That said, if you are more flexible, then this topic might not be important enough to discuss, and you should direct your questions to things that matter to you.

Questions to Ask as a Lead or Manager

As a lead who cares about technical debt, your main task is to find out if you are okay with the company leadership's attitude towards it. Secondarily, you want to make sure that you have autonomy to address tech debt within the guidelines they set. Part 4, "Leadership Practices," describes the kind of environment I'd be seeking. You could use that as a place to start.

Your job as a lead or manager is not to actually fix the debt, but to create the conditions that allow for it to be fixed. During the interview I would also explore how bad the company's debt problems are, whether they are a top priority, and the constraints that you'd be operating under. These are things you can ask directly.

Safety first

At the time of this writing, mid 2025, the job market for engineers is rough. Not horrible, but not great. This comes on the heels of a relatively fantastic stretch.

Whatever the case, before bringing up tech debt during an interview you should consider your specific situation. If you have been out of work for a while and are going to be fine with whatever tech debt you find in your new job, then there's no reason to bring it up. Tech debt is often a sensitive subject with a lot of nuances (as I have discovered in the process of writing this book).

If you don't care that much about it, it's not worth risking an offer. Only you know how picky you can afford to be and whether it even matters to you.

What If You're the Interviewer?

If you're the person conducting the job interview, tech debt is a good topic to cover. Whenever I was interviewing, I often gave a frank assessment of our debt to applicants, so they'd understand the state of our codebase and my feelings towards it. Their reactions revealed a lot.

Most interviewees will have their own set of questions for you, and many of them could set you up to talk about debt. For example, it's common to be asked what your favorite and least favorite aspects of the company are. It should be easy to steer your answer towards tech debt if that's important to you.

Once you're on the topic, you'll have an opportunity to ask your own questions. I would ask a prospective engineer to describe a time when they had to deal with tech debt to solve a problem. I don't think there are right or wrong answers here; my main concern would be if they hadn't given it much thought (depending on how senior they were). I would also want to make sure that if they held any strong beliefs on tech debt, they were compatible with your team's strategy, whatever it is.

I never expected perfect alignment, but if the person I was interviewing's values were drastically different than mine, I would want to discuss it further so the engineer wouldn't be completely frustrated if they joined my team. A diversity of opinion is important, and we should accept that any addition to a team will necessarily change the dynamic in some way. But it's still possible for a misalignment to be so large it cannot be overcome.

When I was at Trello, as part of our first tech interview we had the applicant alter code that intentionally had some debt in it that was similar to the kind of debt in our codebase. We asked them what they thought of the code and checked to see if they identified the debt. We then asked them to fix it. Not only was this a lot closer to the work we did (unlike having them reverse a linked list), but it also gave us a chance to talk with them about the culture of our team. Whether we wound up working with them or not, I appreciated the opportunity to speak honestly about the state of our codebase. I assume they did too.

CHAPTER 16

Fix Tech Debt as Part of Onboarding

A lot of tech teams try to get new hires to get code merged on their first day. I like this idea when the goal is to train them on the procedures for opening a PR, getting a local build working, running tests, and navigating your issue tracker. But for them to be able to do these things, you need good entry-point tickets that can be implemented by someone who doesn't know your codebase.

Paying tech debt is a good candidate for this kind of training, since you can use your tech-debt metrics (see Chapter 9) to identify the best areas for them to work on. Adding unit tests to high value but under-tested areas has no risk of regression, so they won't be able to mess things up. Any work they do should also be trivial to review, so it'll close the feedback loop quickly, helping them meet their merge goal.

This process will also help prep them to tackle their first feature. If you have a small one that you'd like them to take on, first have them learn that area of the code by doing some helpful refactorings to make the code clearer and easier to change. They'll already have to learn that code, so it'll only add a little more time to reflect that learning in commits. When we review the code, we can verify that they understand the area and our conventions because they have to choose names for the functions they are creating.

Finally, having them clean up old code during onboarding sends a strong signal about your team's commitment to paying tech debt and the level of professionalism you expect. If you want to make sure they understand your team culture, it's best to immerse them in it right away.

Onboarding via Tech Debt

I was exposed to this way of onboarding at my first job out of college, when I worked for a company that made a DOS desktop application for pricing foreign exchange options. This domain was totally new to me, so I didn't know what most of the code was even supposed to do.

But every program has non-domain specific code, and in the early 90's, DOS programs usually had their own Text UI screen rendering system, which I could understand on day one. If I had been asked to fix an options pricing bug as my first task, I would have had to spend weeks learning the domain, but instead I was onboarded in an area that was related to my coursework and student projects. The code worked, but it had some debt that needed to be paid—the perfect entry-point task for me.

Our rendering code was very memory inefficient, but that could be fixed in a straightforward albeit tedious way. We joked that these monotonous tasks were character building, but it's the kind of thing that junior developers can (and should) do. I spent my first couple of weeks rewriting each "window" of the application and got to see every screen of the system. This project helped onboard me to the software, its structure, its build, and our issue tracking, as well as version control workflows. Along the way, I paid down some debt incidentally.

This was way before ubiquitous code reviewing was an industry norm, but I was committing small bits every day, and my manager was looking at them and sending me comments. So, I was learning the team culture and style as I generated commits which elicited comments.

I did this again when I started at Trello. I had been a user for three years, so I knew what the app did, but I was still working with a totally new codebase, team, and culture. My first project was to add internationalization (i18n) support. The first step of that was getting each string in the code into a strings file that could be sent to translators. A tedious and character-building task, but I could do it without needing any guidance. My goal was to fix the debt of hardcoded strings, which taught me a lot about the codebase and our process as I did it.

The trail of commits I made helped me remember what I had learned. Each time I looked them over, I was reminded of the areas I had touched, so I didn't need to memorize everything. If I had just read the code and not changed it, my cognitive load would have been constantly increasing. Keeping that load low helped me learn the codebase faster.

Onboarding PRs Need Extra Scrutiny

As you onboard, be sure to keep all of the caveats about debt payment we've discussed in mind. For example, make sure that you are paying debt in areas that are likely to need updates down the road. Paying debt on code that works and won't change any more is a waste of time. We want our new hire to be doing work we value.

It's also important that the new employee's PRs get extra scrutiny during this time. In debt-laden code, it's sometimes unclear what all the dependencies are and what changes might result in unexpected problems. During onboarding and for their first few months, new hires should get more thorough reviews than might be your norm.

Finally, since this code might not have a ticket written already, make sure to write one up. Most likely, you'll need to explain to QA exactly what areas might have been affected and what differences they should expect to see. A lot of times, debt fixes result in no visible changes, so if that was your intention, let them know.

The goal of onboarding should be to help new hires get used to the tools and codebase they will be working with. Paying tech debt offers a perfect opportunity to do this. The changes they will need to make (tests, comments, and refactors) are relatively easy, so they will likely get through several PRs faster than they would if they had to write more complex code or debug thorny issues. Their feedback loops will be shorter. Reviews will be happening more frequently. It's hard to get off-course.

If it's done the right way, paying tech debt lets the new hire work on valuable tasks and helps teach them your team's culture and norms. Even better, because deep domain or codebase knowledge isn't necessary, it makes the onboarding go faster.

CHAPTER 17

Borrow With a Short Grace Period

When I was at my first job, there was a vestigial TODO in a complex area of the code that I kept running into:

```
/* TODO: Is this right? -AW */
```

I knew who AW was, but he had left before I started, so I couldn't ask him what this meant. I wanted to delete it, but I could never figure out if the code it referred to was right or not. I left the company before resolving it, so for all I know, the comment is still there. This was a bad way to use TODO.

To avoid creating a similar problem for my coworkers, for my last PR at Atlassian I searched for every TODO that I had left in the Trello iOS codebase. This was possible because, like AW, we always signed them. I resolved each one in some way and then removed it. Saying "toodle-oo" by removing "TODO: Lou" from our code made me smile.

Looking back on this, I can say that none of the TODOs I had put in were there for a good reason. It's not that they weren't true: it's that leaving them in the code wasn't going to get them resolved. Either they didn't need to be resolved (and should have been removed), they urgently needed to be resolved, or they should have been put in our backlog to be prioritized with everything else. The fact that they were years old, sitting there littering the code and unresolved, was the worst possible scenario.

Since I'm 100 percent in charge of my code style guide these days, I don't allow TODOs to be merged (they won't pass linting). But I do use TODO in the code to reduce cognitive load while I am building a PR. Because I'll be forced to remove it before merge, I can be sure I won't forget them. This lets my brain stop worrying about whatever problem I found.

Another TODO I'm okay with is one that's going to be resolved very soon. I actually prefer it over a ticket when I know the exact line of code that needs to be changed, and I want the message right next to it. If I need the ticket anyway, I'll cross-reference them. I don't commit them. They are there to remind me as I start each new branch, egging me on to do something about it.

Delaying Work to Stay in the Zone

These TODOs help me stay in the zone when I'm coding and not get derailed. For example, I've recently been working on a new web app. The backend is a GraphQL API on top of a database. Each entry point unpacks the request and then calls a function that queries or mutates the database. There are also some complex mutations involving updates, inserts, and deletions that all need to succeed or fail together.

I know that I'm supposed to do this combination inside a transaction, but, like I said, I had just started the app, and I was using an ORM that I'm not very familiar with. I really didn't want to stop my momentum to learn how to do everything by the book. So, I decided to pay lip-service to transactions, wrapping some functions in them to remind myself I needed to think about them but mostly just ignoring them. I would never do this in a production app, but this was an MVP, so it fine. I dropped in some TODOs and moved on. I couldn't merge them because they wouldn't pass CI, but they were constantly in my face in diffs and lint warnings, which was the point.

After MVP, I added a bunch of minor things while my partner onboarded some trial users. Eventually I tried to implement a feature where the API call could result in many inserts and deletions. I could see in tests that I could cause this to fail in ways that corrupted the data.

The result was that I had to stop, learn how to do transactions in this ORM, and then implement them in my data code. It took a few hours, but now the debt got paid off.

I do this kind of short-term borrowing and paying of debt all of the time. I am essentially borrowing to keep myself in flow. I try to keep the debt top of mind by making it very evident in the code. It's like using a credit card where you pay it off within the grace period.

I also make short-lived TODOs whenever I have to figure out a non-trivial algorithm. To make some progress and to ensure I understand what the algorithm is supposed to do, I might code it in a way I know isn't performant but is easy to see if it works. In my MVP, this is fine to deploy because I don't have a lot of data; in practice, there isn't a difference between $O(\log n)$ and $O(n)$ for me at this stage. My partner might be able to show this feature now or give me feedback, but only if I deploy the suboptimal version. When I update the code, I'll remove the TODOs. The old (less performant) code will become the basis of my unit tests, so the work wasn't wasted.

CHAPTER 18

Turn Documents into Code

When I defined tech debt as anything that resists our progress as we try to swim through our projects, I wasn't limiting it to just code. You can find debt in plenty of our other artifacts. Documentation, for instance, can often slow us down. Lots of documents should have been code in the first place.

Sometimes we create documentation because we have to explain something that we can't fix. Sometimes we document software so someone can understand what it does without having to read code. In both cases, though, the source of truth is the code, which means the document is redundant. To make matters worse, documents like this go out of date. Ultimately it would be better if the document didn't exist or was generated from the code itself.

There are three types of documents that seem to suffer the most from this problem: process documentation, specifications, and diagrams. Let's take each one by one.

Process Documentation

Long process documents often (unintentionally) offer a good description of the tech debt in a system because they are written to compensate for it. Here's an example. When I was an iOS developer at Trello, I had to spin up a local server environment every few months to debug something or work with a new server branch. It wasn't fully automated. So, when I first started working on the app, I had to follow a manual process with steps that were always a little out of date.

Over time, the server team made the process better, which slowly reduced the length of the document. The end result was a one-click deploy of any server branch to a staging environment, which meant I never had to run a local dev environment again. The documentation, which had been long and sometimes incorrect, didn't need to exist anymore.

There are two kinds of process documents that shouldn't be automated, though: ones you rarely use and ones that rely on human judgment.

Things that you don't do often aren't good candidates for automation. For those kinds of manual processes, it's more likely that you need even more documentation because it's likely you will have forgotten all of the caveats and context. If it's very rarely used, the person reading it may know nothing. Onboarding documentation is often like this. That's why I always ask new hires to fix the onboarding documents as they go through them—mostly by adding clarifying notes.

Even with AI (maybe especially with AI), there are some things that need human intervention. Even in 1979, IBM's training manual warned, "A computer can never be held accountable, therefore a computer must never make a management decision."[29] I adopt this mindset whenever a script could cause permanent data destruction.

Specifications

Specifications, such as a product requirements document, are useful for figuring that a project is worth doing. We write them

---

29 https://www.ibm.com/think/insights/ai-decision-making-where-do-businesses-draw-the-line

before there's any code, and maybe before we even know if there ever will be any code. At first, it's not hard to maintain the specification because it doesn't need to match anything. But as soon as the project begins, those specifications quickly go out of date.

Does it matter? To answer this, it's useful to think about what you will get out of the specification once the project begins. For me, it does two things: (1) tell developers what the software should do and (2) tell others what the software actually does.

If the document is describing what the software should do and the software *doesn't* do that, that's not a document maintenance issue—it's a bug. So, in that case, there isn't a document debt issue. But sometimes the code doesn't match the specification because we changed our mind and didn't update the document. That causes trouble if the specification is being used after the code is written, which is likely because non-programmers can only read the specification, not the code. For example, it would be a waste of time if QA used an out-of-date specification to write their test plan. The feedback loop would only be closed when developers were assigned bugs because the (old and wrong) specification was not being followed.

To fix this issue, consider creating a companion document to the specification that focuses primarily on the software's actual behavior. It may be necessary to keep the original for historical purposes, but it should reference this new document as the current source of truth. Since the whole point of this new document is to match the code, it would be best if it was either code itself, generated from code, or could be used to generate code.

Using docstrings is one good way to generate API documents from comments that are close to the code it describes. They could still go out of date, but at least you have a better chance of avoiding this than if the document had been created manually. At Atalasoft, where we made SDKs to manipulate digital photos, we maintained a repository of example images to drive our tests.

But we could also use the image database to generate before and after photos to document what our product actually did. When we changed our code, the document was automatically updated.

You can also often use a spreadsheet to generate code. At Trello, our PMs maintained a spreadsheet with the analytics they expected the app to produce, including all of their parameters. I wrote a code generator that would create a Swift function that matched a line in the spreadsheet, and I wrote tests to dump a CSV from the spreadsheet and compare that to what the app produced. The spreadsheet was the specification, and we were always sure that the code matched it and produced analytics in the form our PMs expected.

Diagrams

If you manually draw diagrams to describe your system, they'll become a constant source of maintenance problems. They may look great, and when they're correct, drawings are better than any alternative. But in my experience, it's not worth it. If you want to have diagrams that are always up to date, you should make them with code and keep them simple.

When I do this, I have two requirements:

1. The diagram's code must be in the document, not stored elsewhere and used to generate an image that you manually embed in the document.
2. The diagram doesn't need to look perfect, but it can't be wrong.

Diagrams are usually embedded in some other documentation so that you can read the text and see the diagram at the same time. But if you can edit the document, you need to be able to edit the diagram in the same tool. This means you should be able to edit the diagram's code in the document editor, not in some other tool.

The one exception is if your diagrams are made from your product's source code. In that case, it's okay if you can't edit that inline in your document editor. What matters is that the embedded diagram updates automatically when the code changes. No manual steps.

I won't lie to you: diagrams generated from code don't look great. But they're worth it to me because they are always correct. That being the case, I strongly advise you keep them simple; resist fighting with the layout engine.

If you can't resist, I'll just mention that most of these tools will only let you get exactly what you want if you know the diagramming language at an expert level. Make sure complex diagrams are commented well. You shouldn't be editing them often, and you don't want to make this another source of debt.

For most of my project documentation I use Confluence, which has a plugin for PlantUML, my favorite diagrams-as-code tool. Don't let the name mislead you: it was originally for UML, but it can make lots of other types of diagrams. There are several factors that make this the best diagram tool for me:

- **I'm documenting the system, not designing it.** After I write the code, I'm not trying to figure out how the system should work. I'm just trying to communicate it quickly to help the reader get a quick sense of it. If I were still designing a system I wouldn't use this; I would probably want to draw instead, but honestly, it'd be on paper or with freeform drawing tools, not a diagram drawing tool.
- **I'm okay with the default layout engine.** In PlantUML, it's best if you let the basic layout engine do its thing.

- **I'm comfortable with programming languages.** PlantUML is a weird programming language with syntax affordances for diagramming and other embedded mini-languages, but I can handle it. You shouldn't expect non-programmers to be able to edit these diagrams, however.
- **I like the idea of version control and diffing.** The PlantUML diagram is made with text, so it works well with diffing.
- **I could generate the code.** I hand-code most of my diagrams, but I could generate it if I wanted to. For example: there are tools that can turn deployment configuration files (like Terraform) into diagrams. Code generation LLMs know PlantUML syntax well, so that also helps.
- **I can edit quickly.** If I want to fix a typo in a diagram using PlantUML, I don't have to click-click-click to get to the part where I can type. Also, I can easily copy and paste from other diagrams and samples.
- **I can make my own libraries.** Drawing tools let you make template shapes, but PlantUML is a programming language (with conditionals, loops, math, etc.). The tools it has for abstraction are way beyond what drawing tools typically can do.

I also use Mermaid[30] when my documents are Markdown files in my repository, because GitHub will display them. There are VSCode plugins that will show them if they are embedded in your code.[31] Unfortunately, Mermaid isn't as full featured as

---

30 https://mermaid.js.org

31 https://marketplace.visualstudio.com/items?itemName=
    vstirbu.vscode-mermaid-preview

PlantUML, but the implementation is in JavaScript, so it's easy to embed in web applications (like GitHub or VSCode). PlantUML currently requires running a Java server to embed in other software, but a JavaScript port exists,[32] so maybe GitHub will support it at some point.

Don't Forget Comments that Should Have Been Better Code

I've found that a lot of comments I come across would be unnecessary if the code was better. So, if you are about to write a comment to explain what a block of code does, consider fixing the code instead. And when you see an out-of-date comment, that may be a sign it's time to pay off a debt by making the code clearer (and getting rid of a comment).

The most common example I see of this is when there are subheading-type comments in a giant function, as if it's okay to write a 400-line function as long as it's broken up with comments. I'd much rather those subheadings (and the code beneath them) were in functions, with the original function reading as a high-level outline.

I know it's easier for me to say than it is to do because there are usually variables strewn throughout the code that are coupling the parts of the function together. There are documented techniques to do it, though. A big part of *Refactoring* by Martin Fowler focuses on the untangling you do just inside of a function so that you can break it apart. Much of the book's advice is counterintuitive, so it's worth reading even if you think you know how to refactor. Additionally, his online catalog[33] is really only useful if you understand his reasoning.

---

32 https://github.com/plantuml/plantuml.js?tab=readme-ov-file
33 https://refactoring.com/catalog/

As an example, it's often easier to refactor if you remove temporary variables. You might have put them in as a premature optimization, but in doing so, you coupled two parts of a giant function, making them harder to separate. If an optimization was necessary, you could always reintroduce the variable and pass it as a parameter to a well-structured set of simple functions.

This may feel like a lot of work to fix an out-of-date comment, but the comment was only the symptom of the technical debt. The cause was the complex function.

When Documents and Code Don't Match, One of Them is Debt

In each of these cases, the problem is that you are trying to explain the same thing in two separate ways, one for the compiler and the programmers (the code) and the other one for everyone else. A debt is created whenever one is updated but the other one isn't.

The best way to stop that from happening is to get rid of one of them, which you can do when the documentation only exists because a process is only partially automated.

Another technique is to generate one of them from the other. A spreadsheet is often a good input to a code generation script, and docstrings let us go the other way and generate API docs from our code comments. Look for other opportunities to do this kind of thing. I have found it comes up a lot.

As I've pointed out several times but will point out once again, comments describing complex code often wouldn't need to be there if the code was readable. So instead of rewriting an out-of-date comment, rewrite the code so that you can remove the comment.

CHAPTER 19

End with Broken Code

One way to make sure you end up with testable code and tests that drive it is to practice Test-Driven Development (TDD). When you write code using TDD, you write the test first, which, of course, will fail because the code it tests doesn't even exist. It won't compile. That's okay, because the next thing to do is to write the literal minimum amount of code that will pass the test.

Practitioners of TDD build up clean code by moving back and forth between using a test to specify the next step and then implementing that step. I find that when I do this, it gets me into flow quickly, because if I get stuck, I just write a test that expresses a small next step. But I don't write all of my code this way. When I do the code is likely to be simple and have high test coverage, which makes it easier to change in the future (and therefore has less debt). Avoiding debt is great, but I've also adapted TDD to help clean up old messes.

Here's how.

One common TDD practice is called Red, Green, Refactor. In this practice, first you write a test that fails, then you write code that makes the test pass, and finally, you refactor that code to be clean. (The last step becomes safer to do because you know you have a test covering the code.) The hope is that when you're done, you have code that is both simple and covered by tests.

When I'm working with old code, I like to riff on this slightly with a process I call Green, Refactor, Red. When I start a PR, the tests will all be green, and you would normally want to keep it that way, but if I'm not finished with the PR at the end of the day, I try to end the session with the tests being red. This makes it very easy to pick up the code later because it's broken in a way that so obviously reminds me where I left off.

When I am almost done for the day, I start a refactor and leave the code uncompilable. For example, I might extract some code into a new function but stop before passing in the parameters. I have a bunch of undefined variables in the new function. To compound the problem, I write a test for the new function, also passing in no parameters. That leaves the build red, which is when I stop.

When I get back to my IDE, the errors tell me what to do:

1. Add the parameters to the new function
2. Pass in the correct values at the call site
3. Pass in test values in the test
4. Write more tests to cover the edge cases
5. Commit my work

Doing that work puts me in a coding flow and gives me momentum for the next step of the project.

Target Code that Doesn't Break When It Should

If I don't know what code to break, I do *mutation testing*. I learned this a long time ago from a testing book, though I don't remember which one. Anyway, when I am looking at big block of obtuse code that I think isn't tested well, I just change something small in a way that is obviously wrong. For example, I change a less-than to a greater-than, an AND to an OR, or slightly alter an arithmetic expression. I pick something small that keeps the code syntactically correct. *Now* I run my tests, because even if this code is technically "covered," that's only part of the story;

you also need a way to make sure you have assertions for all of those paths.

If the tests go red, I undo the change because the "bug" I introduced was caught (and I was wrong about it not being tested). But if the tests are green, my goal is to write a new test that fails now but will succeed if I undo the change. It will be a lot easier to write this test if the line of code I changed is in its own function, so I start a refactor and leave it red to leave me something to fix next time.

I find that the practice of leaving code broken is more useful the more time there is between my coding sessions, i.e., I pick things back up after a weekend as opposed to after lunch. But if the gap is going to be too long it probably makes more sense to just finish up the PR and push it, so I probably won't hold onto broken code just before I go on vacation.

We've Come Full Circle

I recently learned that the technique of ending a day's work mid-thought is called the Hemingway Trick. It was coined in response to a 1964 interview he did with *Writer's Digest*, which included this line: "The best way is always to stop when you are going good. If you do that you'll never be stuck."[34]

In the first chapter of this part, "Start with Tech Debt," I suggest starting a coding session by fixing debt. It's a task that's easy to do and gets you warmed up before you transition to your work at hand. In the practice I describe in this chapter, we are just pre-picking the debt to fix next time. Doing this helps us connect the flow we have right now to our next session, helping us get started even faster. I'd like to think Hemingway would approve.

---

34 https://quoteinvestigator.com/2019/01/30/fishing/

PART THREE
TEAM PRACTICES

Part Two showed ways to clear debt as you code. This works by prolonging flow, shortening feedback loops, and reducing cognitive load. Done well, you end up finishing features sooner than you would have by ignoring the debt. Your codebase improves, but it takes time.

Some problems can't wait for that. To fix complex problems faster, we need a plan and enough time to focus on it. We'll need sign-off. These problems can't be fixed on our own or without management support. And when we're on a team, no matter how well we each swim alone, we can't get as far as when we swim together.

So, in this part, I'll describe how I approach bigger problems. To start, the first two chapters explain how to make time to fix tech debt, allocate that time, and agree on our general direction. Then, we'll be swimming together. But getting to our destination will need us to make some choices. We'll need to analyze and compare the costs, benefits, and risks. The final four chapters show how to manage a tech debt backlog and design projects to address it.

I have to warn you: there will be some meetings. To make that easier, I put sample meeting agendas in Appendices 2-4. Once you get going, it will only take a short quarterly meeting to maintain the tech debt backlog.

CHAPTER 20

Make and Spend a Budget to Pay Back Tech Debt

In Chapter 1, I told the story about how when I was under stress to ship an immovable roadmap on time, I completely ignored tech debt. I learned the hard way that that doesn't work. I saw that devoting some time to tech debt can actually be used to deliver faster. That wasn't the only benefit.

My team gained confidence that debt was temporary. It freed them to design solutions that were intentionally short-term focused when it made sense for the business. We didn't always know if a feature would be popular or how to scale it. In either case delivering is how you learn. This isn't "taking a shortcut." It's designing the correct system under current business constraints.

Of course, this won't work if you never revisit these choices. The team will become frustrated, and eventually they'll learn not to propose any short-term solutions to a problem even when you need them. You deployed fast to learn fast. Now you need to use what you learned. Regular tech debt management is what makes this process work, so you should allocate a budget ahead of time to show that you will do it.

Give Engineers Some Budget to Allocate Autonomously

At Atlassian, our team divided our budget into two buckets. We called them **product-led** (or **PM-led**) time and **engineering-led** time. A product-led initiative was something that ultimately rolled up to a business objective. The requirements came from the Product Manager (PM), who would define acceptability. Anything that improved the software for customers was likely product-led. An engineering-led initiative was something where the specs would only be understood by an engineer and whose benefits were internal to the team. It included tech debt and more.

There are other things we do for ourselves that don't pay debt. For example, regular dependency updating, build machine maintenance, or developer tooling. Any internal task which our PM didn't specify or care about was also classified as engineering-led. But the bulk was tech debt.

These budgets were applied to projects and stories. It was understood that there was also overhead. We didn't try to classify every hour or every meeting, for example. A high-level understanding of how our work was split between the two was enough.

Allocating the Budget

Our Engineering Manager (EM) and PM negotiated a figure for the year. For you, the split would be based on where your software is in its lifecycle. Part 4 ("Leadership Practices") explains this more. Generally, favor product-led projects when you're seeking product-market fit. Do less of that when the product is too debt-laden to make progress. Balance the two when you're in between.

The division of time may vary each month. This is fine. It just needs to match over the entire year. This happened to us if we had an emergency or if our designers needed time to prepare work. We'd do extra engineering-led work and make it up later. To make sure you have an accurate accounting of what's really happening, it helps to have a process to split and track the time.

For example, if you were aiming for 20 percent engineering-led time, there are a few ways to do that:

1. Plan your sprints so that 20 percent of the story points are engineering-led.
2. Devote every Friday (20 percent of the week) to paying down technical debt.
3. Devote every fifth sprint to paying down technical debt.
4. Assign 20 percent of your team to paying down technical debt (rotating who this is every quarter or so).

I've done some variation of all of these. In my experience #4 works best because of how we review and reward developers. Your review is about your big projects. Something that is 20 percent of your work isn't going to be discussed. It isn't a priority. That's why the first three ways don't work as well. But, in the fourth allocation method, 20 percent of the team spends all of their time on engineering-led problems. If you are supposed to spend 100 percent of your time on something, you will talk about it in your 1:1 with your manager. Your manager will remember it. It will be in your review.

Additionally, making paying tech debt someone's full-time job for a quarter lets them plan bigger projects. It's hard to get a PR done in one day. If you only work on a project on Fridays, it takes month to get anything deployed. When you work on it full-time, however, you can deploy much more frequently. You can plan for the future. For example, I would start by deploying some extra monitoring that could measure impact or catch regressions.

Working on something every day gets developers into the zone faster. You need this for giant restructuring/rewrite slogs. If you only get one day a week, you have to reacquaint yourself with anything big. You spend that one day quickly. You already know how I feel about flow and the cognitive load associated with context-switching.

Assigning debt to an individual will also make it more likely that this debt paydown isn't spread out. Code changes tend to cluster when a single developer is doing a larger project. This makes it easier to check that it hasn't caused regressions. If you just allowed in a bunch of random stories, the changes might not cluster in a way that makes it easy for QA to test.

Finally, this kind of allocation makes it easier for managers. They just assign developers to each bucket. They don't have to monitor individual stories over time and can adjust to temporary ebbs and flows.

Whatever you do, the important thing is that you spend the budget. This is what gives the team confidence that they can fix the problems that are slowing them down. Once a system is in place, they can build feature implementations that are appropriate to the current business objectives, knowing that they could cause problems later. They'll have data to guide them. Then, using their own discretion, they can decide when and how to fix those problems.

CHAPTER 21

Add Tech Debt Rules to Your Style Guide

In the Personal Practices part of this book, I advised you to move your code in noncontroversial ways. We're doing this to avoid long code review cycles. If you are randomly changing function names just because of personal preference or rewriting working imperative loops as functional maps, and your team isn't fully onboard, then you've wasted time doing it. You're going to waste even more time during code review. If your whole team does this, the result will be chaos.

Questionable changes invite questions. If you get a lot of subjective comments, start by considering whether your work aligns with the team's norms. In Chapter 4, when discussing feedback loops, I mentioned that the review comment may be telling you that you need to improve. Your goal should be to avoid rejection by doing work that is correct and easy to verify.

But sometimes, it's just nitpicking. The most frustrating code reviews of my work have been ones where the comments were all just undocumented style preferences. Each change request was presented as if it was self-evidently better. The best comments point out something objective. For example, a bug. Your goal as a reviewer, should be to mostly point out objectively unacceptable code.

Of course, there are subjective aspects too. That's fine, but we need to agree on the choices and write them down in a style guide. When you do that, you are creating an objective standard. Style comments need to cite it instead of personal opinion.

Use style guides to settle disagreements. After a rule is written, it's okay for a review comment to point out how some code misaligns with the guide. Those are the only style-related comments that should be allowed. If there are new disagreements, resolve them outside of the review. Add them to the guide, but don't stop there.

In Chapter 18 ("Turn Documents into Code"), I said that documents become tech debt because they are hard to maintain and remember. The same thing goes for a style guide. It's better if it's code instead.

Most team style guides already have all of the simple things automated. Issues like tabs vs. spaces, brace style, and line-length can easily be checked by linters, formatters, and commit/merge rules. Use these tools to avoid long reviews. Remember that short feedback cycles drive developer productivity, so any action you can take to reduce review time will help you deploy your change faster.

Tech Debt Rules Can Be Automated Too

The tech debt rules in your guide are harder to automate than formatting rules. It is possible, though. I showed an example in Chapter 9 ("Make the Effects of Tech Debt More Visible"). On my projects, I have a rule to never change the behavior of a function if it has a high "change risk," which means that it is both complex and undertested. I use plugins in my editor to highlight code like this. So, if I need to change it, I must first either refactor it to lower its complexity or test it to increase its coverage.

The tools nudge me to refactor and test the code. I do that first, and then I make my intended change. This not only lowers my own cognitive load but also helps the code reviewer understand the code. They see code get simplified, then tested, and finally, changed to implement the new behavior. Each commit is easy to verify.

The code reviewer (who agreed to this rule) expects these initial commits. They use them to finish the review faster. If I committed changes to complex functions, our guide would make it valid for the reviewer to ask me to make the code clearer. Our rule means that this is not being nitpicky.

I also have a test coverage threshold for the code that changes in a PR, which I find more effective than enforcing a threshold on the entire project. I use a tool called diff-cover[35] to make that easy. Diff-cover uses standard test coverage files as a starting point, but it filters them down based on code that changed in your branch. I can set a high coverage threshold because I only need to test the changed lines.

There are other tools that can automate a tech-debt rule. These include:

- dependency maintenance tools like Dependabot on GitHub
- your compiler's warnings
- security-focused static analysis tools (see the guide maintained by OWASP)[36]
- accessibility static analysis tools (see the guide maintained by W3C)[37]

---

35 https://github.com/Bachmann1234/diff_cover
36 https://owasp.org/www-project-devsecops-guideline/latest/
    02a-Static-Application-Security-Testing
37 https://www.w3.org/WAI/test-evaluate/tools/list/

If you have a rule that is causing round trips in code reviews, consider building your own tool to try to find it. Run that tool locally before you open a PR.

Code reviews are just one feedback loop to shorten. There are more. In codebases that need to be internationalized, I automate a check that every string has a comment describing its use in the application. Without that, I'm sure to get a question from a translator. If you have design reviews, consider automating adherence to the design system. Any wait for an asynchronous check by another person might result in a rejection. It's much better to get that rejection immediately from a tool if you can.

A Target Architecture

Rules work for new code. Long-lived codebases sometimes have sections of code that we wouldn't accept now, but that we can't easily fix. The problems often go beyond simple style issues. As codebases age, programming languages change, new paradigms become popular, and we learn more about how to structure our software.

If you want to make steady progress towards fixing old code, you need to make sure the team has a common understanding. So, maintain a document that describes the ideal architecture for your system given your current understanding. You might fix it all one day. But simply having it helps because there will constantly be opportunities for developers to use this target to justify smaller steps.

A few years after Swift came out, it was clear to us on the Trello iOS team that it would be better if the app was in Swift, but we couldn't drop everything and rewrite the app. We did agree that all new code should be in Swift, however. The interoperability between Swift and Objective-C made it easy to transition one file at a time. So, we also agreed to convert files before making major updates to them.

We had a similar philosophy when it came to UI code. Our code implemented UI in various ways. Sometimes we used Interface Builder (IB), and sometimes not. At some point we agreed that we preferred Model-View-ViewModel (MVVM) without IB to other UI architectures. We specifically eschewed other possible architectures, like VIPER or MVC. Just like with Swift, we allowed conversions of single views to MVVM when they were being updated for other reasons. Over time, the code became more homogenous. It was easier to transfer knowledge as we moved around in it.

We felt the benefit of these decisions immediately in reduced communication. Engineers didn't have to wonder if they could update an area of code they were about to change. They didn't need to get permission from a lead. In a review, it wouldn't be a topic of discussion.

It also made code easier to review. In the case of my Trello iOS team, because we started a PR by first rewriting Objective-C as Swift, we not only got better code, but we also ended up learning it while getting into the flow. Then, when we made the change that was driving the PR, we were able to do it with the confidence that we truly understood the code. That understanding flowed to the reviewer as well.

As we'll see in the next few chapters, agreements drive tech debt work. Having a target architecture is important because bigger agreements drive bigger payments.

CHAPTER 22

Schedule Regular Tech Debt Meetings

Tech debt is one of those things that an engineer can fixate on. I've felt it. There have been many days when I've come out of a coding session frustrated and feeling like I've been thwarted by the codebase from getting anything done. I finally give up and go to my next meeting. I can't help but vent. When you're working in a codebase with a lot of debt, every meeting can feel like the right place to bring it up.

One solution is to have regular meetings about tech debt with your team and limit debt-related discussion to just those times. If debt is on the agenda for any other meeting, move it. The next four chapters will go into detail about what I think you should do at those meetings. This chapter is an overview.

Asynchronously, before the meeting, your team should make a tech debt backlog. Then, a big chunk of each tech debt meeting should be dedicated to figuring out the costs and benefits to dealing with (or ignoring) debt items that you have identified. You'll each have intuitions, but they might not match. It's worth spending some time to ground that intuition in evidence.

To do this, I developed a tech debt scoring system to drive the discussion and figure out the costs, benefits, and risks of each option. The effectiveness of objective scoring was described in *Thinking, Fast and Slow* by Daniel Kahneman. He discovered it when evaluating prospective soldiers:

I decided on a procedure in which the interviewers would evaluate several relevant personality traits and score each separately. The final score of fitness for combat duty would be computed according to a standard formula [....] I made up a list of six characteristics that appeared relevant to performance in a combat unit.

[...] I told [the interviewers], "and when you are done, [...] close your eyes, try to imagine the recruit as a soldier, and assign him a score on a scale of 1 to 5."

[...] I learned from this finding a lesson that I have never forgotten: intuition adds value even in the justly derided selection interview, but only after a disciplined collection of objective information and disciplined scoring of separate traits.[38]

We'll get more into the scoring system later.

To start the process off, I recommend two 1-hour meetings, just to understand the breadth of your debt. Schedule them one week apart. I would treat these first two sessions (hereafter called "Kickoff Meeting" and "Kickoff Follow-up") as a separate phase from your regularly scheduled discussions. After you do them, you should have a prioritized backlog of tech debt and a plan for addressing it.

The next meeting will be in a quarter. Between meetings, the team will follow the plan they created. In my experience, quarterly meetings should be enough to keep you on track. The team will keep the backlog up to date with changed scores and new items. Your backlog shouldn't grow much, though. If it does, and you can't wait for the next meeting, just schedule an ad-hoc one.

---

38 Thinking, Fast and Slow by Daniel Kahneman, Chapter 21: Intuitions vs. Formulas

Before the First Kickoff Meeting

As I mentioned, the goal of the Kickoff Meeting and Kickoff Follow-Up is to create and prioritize a backlog of tech debt. This backlog is a place to plan, not execute. That being the case, I recommend keeping your tech debt backlog separate from your normal feature backlog. Nothing in the list has been approved to work on yet.

Your debt discussions are internal. They are not meant to be understood outside of your team, so it would be confusing to mix them into the feature backlog. Think of the tech debt backlog as analogous to what your product manager might use before they decide on a roadmap—for example, in products like ProductBoard or Jira Product Discovery. This is where we consider the options. Once you decide to address a debt, you will assign tasks using your normal work planning tools. That's how we'll coordinate with others.

Pick a place for the tech debt backlog. A shared Google sheet, a table in a Confluence or Notion page, or a Trello board are all good options. If you don't have a strong opinion, use an online spreadsheet. That will make it easier for doing math and visualizing, as we'll cover in a bit. I put a link to a sample Google Sheet in Appendix 1. I'll be using this spreadsheet as the basis for the rest of the chapters in this part of the book.

I want the meetings to be short. To make that happen, encourage your team to add tech debt to the backlog beforehand. Write in one or two obvious ones. Talk about it in 1:1s. If you don't see the list grow, bring it up. To give you an idea of what to list, I filled in a backlog on the sample spreadsheet. It lists a bunch of the tech debt I've had to deal with over my career:

Tech Debt Backlog ∨ 🖩	
Short Name ∨	**Summary** ∨
Dependencies	Dependencies are updated haphazardly and get out of date
Design System	It isn't easy to match the design system
FixMVC	Our UI code isn't consistent or testable
Folder Mess	The code folders don't match Xcode groups
I18n	The app is not internationalizable
Markdown	iOS markdown doesn't 100% match web
Permissions	API permissions are represented by guard-statements, not data
Python3	Update to Python 3
Sockets	Sockets system uses old tech and causes lots of bugs
Sync Reliability	We don't know what's causing our synchronization problems
Unit Tests	Our project has no unit tests
Window Memory	Our windowing system uses too much memory
Zero Warnings	There are too many compiler warnings and we can't see if there are real issues

Each entry should include two things:

1. A short name that you can use to refer to the debt
2. A one-line summary (if you need something longer, just link to it)

The scope of each listed tech debt issue is going to vary wildly. That's okay. It's also okay to host asynchronous discussion on the document. It's important to resolve obvious questions before the first Kickoff Meeting, so if there's an item on the list that you don't understand, ask about it beforehand. That will keep the beginning of the meeting short. Ideally, everyone at the meeting will already know what each item refers to. We want to use most of our (limited) time to have a deeper discussion.

Describe the problem debt is causing. We developers are both the customer who is experiencing the problem and the one who will fix it. We may think we know exactly what to do to fix an issue. But at this point, it's best to keep your customer hat on and try to describe the problem without presupposing a solution.

As much as possible, try to limit descriptions to only describing the current state of the debt, the problems it's causing, or the acceptance criteria for fixing it. Developing an appropriate solution is what your meetings are for. If a solution is obvious, fine—go ahead and mention it. But if it's not, leave it for later.

Give everyone at least a week to add items to the backlog. Make sure they read it too. Use the time to resolve any questions about the items in advance. We'll go over the list at the top of the meeting, but we want that to be quick.

The next four chapters go into depth about each meeting. We'll learn about the pre-work that keeps the meeting short and what we'll do together. But before we dive in, it will be helpful to see an overview of the timeline and a checklist.

To Start:

- Schedule a Kickoff Meeting.
- Schedule a Kickoff Follow-up Meeting one week after the Kickoff.
- Create a place to collaborate on a tech debt backlog.
- Ask the team to add items to the backlog before the meeting.

At the Kickoff Meeting (See Appendix 2 for a sample agenda)

- Resolve open questions about the backlog.

- Stack rank each item against each of the eight dimensions of tech debt (see Chapter 23).
- Assign scoring tasks to the team.

Between the Kickoff Meeting and Kickoff Follow-Up

- The team scores the items (see Chapter 24 and Appendix 5).
- The team prepares evidence for scores that may be controversial or that change the agreed upon ranking.

At the Kickoff Follow-Up (see Appendix 3 for a sample agenda)

- Use the scores to set a priority for discussion.
- Discuss each item in priority order (see Chapter 25).
- Decide on a plan for the next quarter based on the scores and your budget (see Chapter 26 and Appendix 6).
- Schedule regular meetings for every quarter.

Before Each Tech Debt Regular Meeting

- Do the planned work.
- Update the tech debt backlog (new items and scores).

At Each Tech Debt Recurring Meeting (see Appendix 4 for a sample agenda)

- Celebrate the progress since the last meeting.
- Discuss the updates to the backlog since the last meeting (possibly change scores based on new information or work).
- Plan the next quarter's tech debt work.

This process is a starting point. Use your retrospectives to tailor it to your team's style. We want the team to know that debt is temporary. To get the benefits, you need a place to track debt, time to pay it down, and autonomy.

The next few chapters go into the details on the process. This includes a scoring method that helps drive discussions about what to do about your tech debt. The scores let us compare items to each other to prioritize them. They also help surface and re-solve arguments that might otherwise be based on a gut feeling. That feeling might be right. If so, the scores will unpack it.

CHAPTER 23

Rank Debt Items at the Kickoff Meeting

As we just discussed, the purpose of dedicated tech debt meetings is to give the team confidence that your tech debt items will be addressed. We'll start by analyzing the backlog they generated. It's likely that different members of the team will have different perspectives. We need to resolve that. The goal of the meetings is to come to an agreement and assign work.

I use a scoring rubric to drive this discussion. As you may know, this approach is a lot like what product managers do to assess and plan various strategies,[39] so in these next four chapters we're going to apply ideas from product management to help the team come to a mutual understanding. The scores will guide our intuition to make better decisions.

In this chapter I will introduce the eight dimensions of tech debt. We will see how to use them to rank the tech debt backlog items in various ways. This exercise is the main focus of the Kickoff Meeting.

Kickoff Meeting Overview

A sample agenda for this meeting is in Appendix 2, but here's a high-level overview of what happens:

---

39 https://www.atlassian.com/agile/product-management/
    prioritization-framework

Part 1: Complete the backlog document (10-15 minutes)
At the end of this, everyone should understand what is meant by each item.

Part 2: Stack rank the debt along various dimensions (30-40 minutes)
At the end of this part, you will have eight different sorted lists of the items in the backlog.

Part 3: Assign team members to score the dimensions (5 minutes)
At the end of this part, each team member will be assigned to help score some of the lists. The scoring work will be done before the kickoff follow-up meeting.

The Dimensions of Tech Debt

For most of the kickoff meeting, we'll be stack ranking the items in eight different ways. The purpose is to try to drive the discussion towards the objective attributes of the tech debt items. We want to rely less on intuition at the start.

To that end, I propose ranking the items along the following eight dimensions:

1. **Visibility:** If this debt were paid, how visible would it be outside of engineering?
2. **Misalignment:** If this debt were paid, how much more would our code match our engineering values?
3. **Size:** If we knew exactly what to do and there were no coding unknowns at all, how long would the tech debt fix take?
4. **Difficulty:** What is the risk that work on the debt takes longer than represented in the Size score because we won't know how to do it?

5. **Volatility:** How likely is the code to need changes in the near future because of new planned features or high-priority bugs?
6. **Resistance:** How hard is it to change this code if we don't pay the debt?
7. **Regression:** How bad would it be if we introduced new bugs in this code when we try to fix its tech debt?
8. **Uncertainty:** How sure are we that our tech debt fix will deliver the developer productivity benefits we expect?

These dimensions were derived by first considering the various costs and benefits to paying down (or staying with tech debt). This idea comes from Bob Moesta's idea of the four forces that drive progress.[40] In his force diagrams, he puts the status-quo on the left and the progress we want to make on the right. Then, he puts in forces that push and pull you in each direction.

For tech debt it looks like this:

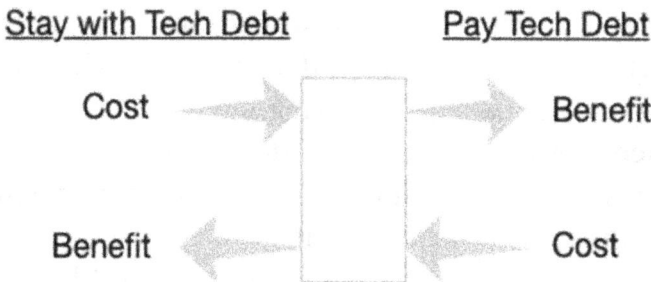

Stay with Tech Debt	Pay Tech Debt
Cost ⟶	⟶ Benefit
Benefit ⟵	⟵ Cost

The cost of staying with tech debt is a force that pushes us to pay it. The benefit of doing so is a force that pulls us to do it. But there are also benefits to doing nothing. They will pull us back. Along with that, the cost to paying tech debt pushes us towards

---

40 Inspired by Bob Moesta: https://jobstobedone.org/radio/
unpacking-the-progress-making-forces-diagram/

staying with the status quo. The relative sizes of these forces net out and guide us on what to do. We'll use the net force to prioritize the list.

To find the values of these forces, I identified two independent drivers for each force based on my experience:

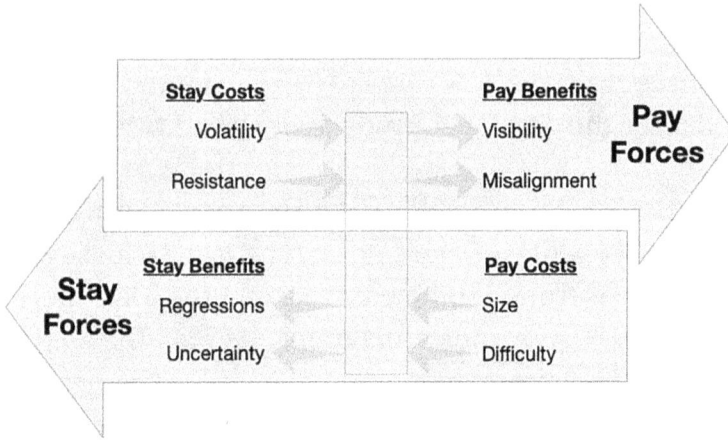

Each pair of dimensions indicates a cost or benefit. For example, when tech debt is highly **visible** to customers or causing a lot of **misalignment** with our documented engineering values, that points to benefits of paying the debt. **Volatility** and **resistance**, however, are costs of doing nothing. If the combined scores of these four dimensions are high for an item, this suggests you should probably pay it.

The stay forces work in the opposite direction. A high chance of **regressions**, high **uncertainty** that our remediation plan will deliver productivity, and a proposed plan with large **size** and high **difficulty** are all forces that suggest we should stay with this debt (or at least tread carefully).

As we shall see, it will rarely be this clear. Usually, it's a combination of high and low forces that need to be considered both on their own and together.

Start by Putting the Items in Order as a Team

After we have made sure we understand the backlog, to get warmed up to the idea of the dimensions, I start by sorting our list against each dimension. So, for example, under a column that says "Visibility," I would ask the team to list each debt item in order from most visible to least visible. When we score this later, it will be easier because we know the numbers should get smaller in the list.

This process should be fairly quick and easy. As I mentioned earlier, be sure to timebox this part of the meeting so you don't spend too much time on each item. We have about five minutes for each. Take notes about any disagreements.

Here's how I approach the process:

1. Pick a dimension.
2. Sort the list of tech debt backlog items along this dimension from highest to lowest.
3. Group items that I think are close to each other, because they will likely get the same score.
4. Repeat this for each dimension.

Let's apply this to my sample backlog (from Chapter 22) along the **misalignment** dimension. The question we're trying to answer with **misalignment** is "If this debt were paid, how much more would our code match our engineering values?" The item associated with the most improvement should be at the top. The items near the bottom may have other good reasons to do them, but becoming more aligned with our values is not one of them.

Let's apply that question to our list. To refresh, here are the names of the thirteen debt items on the list: Dependencies, Design System, FixMVC, Folder Mess, I18n, Markdown, Permissions, Python3, Sockets, Sync Reliability, Unit Tests, Window Memory, Zero Warnings.

The items that most misalign with my team's documented values are **Design System**, **FixMVC**, and **Unit Tests**. For each of those items, our team wrote up a section in our style guide and made a commitment to fix them over time. On the other end of the spectrum, the items named **Markdown, Permissions, Sockets, Sync Reliability**, and **Window Memory** have other issues, but not misalignment. Going through the whole list, I end up with this ordering and grouping, from most misaligned to least:

Design System
FixMVC
Unit Tests
I18n
Zero Warnings
Dependencies
Folder Mess
Python3
Markdown
Permissions
Sockets
Sync Reliability
Window Memory

We repeat this for each dimension, and
meeting we have sorted the list eight different
sort helps ensure that the scores will match th
understanding.

Before the end of the meeting, everyone on
sign up to score dimensions. For example, if yo
members, you could assign one dimension each.
each would have to do two. If you had sixteen, y
up into eight pairs. Everyone should have to sc
dimension. The next chapter will go over the din
tail and explain what the team should do betwee
and the follow-up.

CHAPTER 24

Score Tech Debt Along Cost-Benefit Dimensions

We made a list, and now we need to prioritize it. While it's tempting to do that just off intuition, each item has a different impact and scope and are not easy to compare. It's also likely that team members each have different experiences and feelings about each item, which we need to reconcile. "You can't compare apples and oranges," right? At least those are both fruit. Comparing an undertested module to a UI framework rewrite is like comparing apples and New Jersey.

Tech debt items are multidimensional. Before we allocate resources to address them, we have to analyze their underlying dimensions. That analysis lets us look at the backlog from different perspectives. Then, we can cluster or separate items based on similarity. Finally, we can design plans that are appropriate to each item's value and risk.

The dimensions help us estimate costs and benefits. They reveal what's driving that estimate, which guides our decisions. This is especially useful in identifying debt we shouldn't fix. But if we want to fix it anyway, the dimensions help us do it safely.

Understanding each tech debt dimension is important. So, in this chapter we're going to go into more detail about each. Here's a refresher:

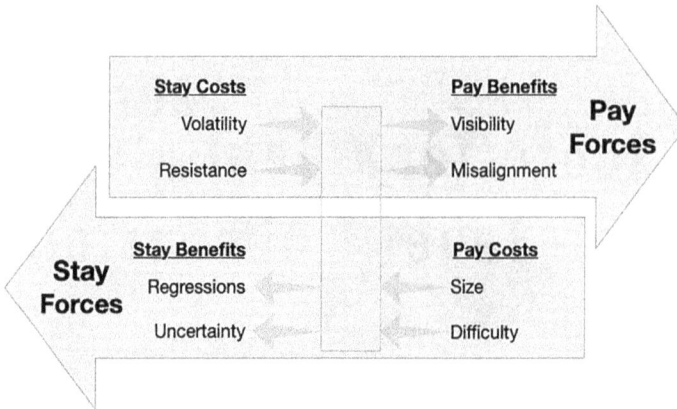

1. **Pay Benefits** are forces pulling you to fix a debt.
   a. **Visibility**: If this debt were paid, how visible would it be outside of engineering?
   b. **Misalignment:** If this debt were paid, how much more would our code match our engineering values?

2. **Pay Costs** are forces pushing you back from fixing a debt.
   a. **Size:** If we knew exactly what to do and there were no coding unknowns at all, how long would the tech debt fix take?
   b. **Difficulty:** What is the risk that work on the debt takes longer than represented in the Size score because we won't know how to do it?

3. **Stay Costs:** are forces pushing you to fix a debt.
   a. **Volatility:** How likely is the code to need changes in the near future because of new planned features or high-priority bugs?
   b. **Resistance:** How hard is it to change this code if we don't pay the debt?

4. **Stay Benefits** are forces pulling you back from fixing a debt.
   a. **Regression**: How bad would it be if we introduced new bugs in this code when we try to fix its tech debt?
   b. **Uncertainty**: How sure are we that our tech debt fix will deliver the developer productivity benefits we expect?

Each dimension will be scored for each tech debt item on a scale of 0-5. High scores indicate a high force. Once we have scored the dimensions, we will calculate a net force, which will give us a general direction (pay or stay) and a magnitude.

The Dimensions of Tech Debt

In this section I'll go over each dimension, give examples, describe how to score it, and list the types of evidence that usually indicate a score should be high. To see a summary, look at the scoring guides in Appendix 5.

I. Visibility

In some codebases, tech debt causes problems that others can see. In 2004, for example, I was hired to help rewrite a system that published data for the electricity market. It had high availability requirements, since energy markets operate 24/7. Although it met that requirement, there were several visible signs that the system was in distress. Our colleagues outside of our team could see it. Our customers, energy traders, could see it.

For one, it was hard for DevOps to diagnose issues because there was no consistent logging. There was also major strain on our databases because our system could not pool connections. This led to us getting calls to help at all hours. To work on our system, you needed to know five programming languages. So, HR had a hard time finding candidates for our team. There were delays producing crucial morning reports. This was a tracked result. Failing to produce this report on time jeopardized everyone's bonuses.

When tech debt is visible, everyone wants it fixed. We don't need to speculate on future productivity benefits, because any short-term gains are often enough to justify the work. You are likely to get the budget you need. It's easier to make sure the project has the time and resources proportionate to its value.

If it is highly visible, there should be evidence of that. You should have support tickets, low survey scores, warnings from security audits, etc. In fact, there's a good chance your PM wants this addressed. It's also in your debt backlog because you want to pay its debt, not work around it.

If it's on your PM's radar, then use their analysis. For example, if they use the RICE framework, then you should use their high Reach (number of customers affected) and Impact (size of the effect) to support a high visibility score.

When you're scoring this dimension, it's tempting to incorporate factors that are visible to your team, like developer experience, to boost the score. But remember that in this scoring system, "visible" means *externally* visible. Try to keep your score consistent with that. Higher visibility scores should indicate that the debt is affecting customers and causing problems for support and sales. Midrange scores should be used when PMs, QA, DevOps, and those closer to engineering feel the issue. The internal factors will be represented in other dimensions. So, when only engineers can see the problem, visibility should be scored low.

2. Misalignment

Visibility helps us get support from others to fix debt because they want the problem fixed. There's an internal benefit to consider too. I recommended that you document your values and ideal architecture. It's useful to consider the impact of any misalignment to it. If addressing an item of tech debt will result in code that is closer to our ideal, then that's a force that pulls us to do it. Code with low misalignment causes less cognitive load. It makes review loops faster. So, if misalignment is high, we should fix it because it's slowing us down.

For example, the Trello iOS team did not have a UI design system at the start, so there was no representation of one in our codebase. Shadows, spacing, and other UI attributes were implemented in different ways. It was hard to make global updates. Later, when a design system was established, we represented it in code and started migrating UIs to it as they were touched. Adherence to this system became a requirement. So, code that was not migrated yet was now misaligned with our standard.

Similarly, we established Model-View-ViewModel (MVVM) as our preferred architecture for new screens. So, it was acceptable to migrate older screens as they needed work. Our team culture encouraged us to fix misaligned code when it was encountered during our regular work. But you couldn't just do whatever you wanted. Fixing misalignment required agreement first.

Misalignment mostly depends on where the standard comes from. The more widely agreed upon the standard, the higher the score should be. Concern over misalignment among the engineering team is less important than a mandate by the CTO. But, if you can't point to any agreement at all, then misalignment should be scored low.

3. Size

The size of a technical debt project is the amount of effort it would take to pay it down if we assume that we won't run into any problems. If we think of technical debt as financial debt, it's the principal payment.

Along with its difficulty, this captures the cost to pay a debt. We are intentionally not including "interest payments" here because they are the cost of *not* paying the debt (and are represented in the volatility and resistance dimensions). The size of a debt correlates to the amount of code that has the debt.

Whatever methods you use to estimate your project sizes would be fine to apply to sizing your technical debt projects. In fact, it might be easier because we already have our existing implementation whose size we can measure directly. For example, if the debt that you are sizing is to convert pages from Angular to React, you can use the number and size of pages and components as a guide to the debt's size.

4. Difficulty

As I just mentioned, difficulty (along with size) is a key measure of the cost of paying down a debt. It captures the risk that our size is wrong.

If we're dealing with messy code that could use some refactoring, tests, and documentation, then it's relatively easy to make those changes without introducing bugs. But if our plan requires a total rewrite, then the difficulty is higher because there are things we might not know until we start.

The degree of difficulty is largely determined by your plan to address the tech debt. Projects that are trivial can be scored low, but otherwise, use the difficulty score to express your confidence in your size estimate.

Here's an example. When I joined Trello, the iOS and web versions each had their own implementation of Markdown, which meant that there were slight differences in how each client rendered a card description. When this became the source of a lot of the Trello iOS support tickets (and open high-priority bugs), I decided to port the web version's JavaScript Markdown parser to Swift line-by-line. The size of the project was just a function of the size of the existing source, which was small.

The risk was similarly small because the work was straightforward. I could lower the risk even more by generating a ton of test cases. If, instead, my plan was to alter our existing Markdown parser to match the web, the risk would have been much higher. I might have had an endless list of slight Markdown rendering differences, and fixing each one could have introduced a new problem. So, the difficulty score you assign depends on your plan.

All of this assumes you can come up with a potential solution. If you don't even have a plan to pay down the debt, then the default difficulty score should be high. To reduce that score, you may need to do a spike that results in a plan with a size and difficulty you can estimate.

5. Volatility

If a system has no visible problems and there are no plans to add any features to it, paying down its debt will be a waste. This code doesn't need to be easy to change. Even worse, you might accidentally introduce new bugs.

Volatility captures how much the code will likely change because it needs new features or has high-priority bugs. We'll have more encounters with this code. If you have high volatility, then you are constantly making interest payments on your debt. It's a force that pushes us to pay it.

As an example, at Atalasoft, we were constantly having problems with our installers. They were made with an older, brittle technology. Unfortunately, our roadmap was going to make us update them frequently. The code was definitely going to change. We'd be feeling the debt. That fact eventually drove us to rewrite the installers using a newer technology that eliminated the technical debt.

When scoring volatility, we care most about short-term future changes. We can estimate that in two ways. The first way is to assume that your short-term future volatility is correlated with your recent past. Your repository history can be used to estimate that.

The second way to predict future volatility is to use your roadmap for the next two quarters. Look at it when you are sorting and scoring volatility. Consider if this debt is going to be encountered soon.

6. Resistance

High volatility isn't a problem if the changes are easy to make. It only tells us how much the debt matters. To get a fuller picture of how much it costs to live with the debt, we need to combine volatility with resistance.

Resistance is the measure of how much the code will resist a change you are trying to make. Code that's highly resistant is hard to understand or risky to change. It might be convoluted, untested, or undocumented. When the code has a lot of bugs (i.e., has high visibility) that are causing us to change it a lot (i.e., has high volatility), those problems can't be ignored.

It's not just poorly written code that has high resistance. The code you're working with might have been fine when it was written. The reality is that lots of tech debt happens because the world changes. Even if your system represents your best ideas of how to solve the problem at hand, your ideas will get better. The problems will change.

I ran into this in the nineties at Astrogamma. We evolved our MS-DOS, text-based application to one that ran on Unix and Windows NT with manageable resistance that was confined to our rendering code. The ports were still single-user, desktop applications, though. But when we needed a web version of the app, the entire codebase resisted the change.

To go back to our financial analogy, resistance is related to the size of the interest payments. The more resistance there is, the higher the payment. Resistance, though, is only the *amount* of the payment; volatility is the frequency of them. Thus, the overall cost of not paying technical debt comes from a combination of resistance and volatility.

I consider code to have higher resistance if it is both complex and undertested. There are several tools that can help you measure this. The current codebase I'm working on is a web-based application written in TypeScript on Node and React, so I use VSCode extensions to show the code's coverage and complexity. (See the section in Chapter 9 "Add Tech Debt Detectors to Your Editor" for more details on this process.)

To measure code coverage, you need automated unit tests. If you have them, most testing frameworks have a way to generate coverage reports. These tools can generate coverage data files that are easy to analyze. If you don't have any tests, I'd score resistance relatively high.

To measure complexity, you can also use a tool that counts branches in the code. There are several ways to do this (e.g., cyclomatic complexity[41]) but the exact algorithm isn't important; we only need to gauge the magnitude of the complexity, not an exact number. It's enough to know if a function has high complexity or not. For example, a function with a lot of nested loops and conditionals is more complex than a three-liner with no branches. As with test coverage, there are plenty of tools that can measure this for popular languages.

---

41 https://en.wikipedia.org/wiki/Cyclomatic_complexity

Looking at code-level metrics works when the overall architecture is right, but the code isn't a good implementation of it. Sometimes the problem is at a macro level, like when my DOS desktop application needed to work on the web. In that case, even if the code was highly tested and well written, it would still resist being a web app.

7. Regression

If your code implements a part of your product that is heavily used and depended upon, then any change to it risks adding new bugs. We call this kind of bug a regression. Leaving working code alone is a benefit of not paying tech debt. It's a force that pulls us back from fixing debt in it.

If you need to add features to this code, then the future volatility will be high, which indicates that tech debt should be addressed. But volatility makes a high risk of regression worse. If your users are highly dependent on a set of features to work correctly, then fixing tech debt might break something, causing a regression. It's a tradeoff you need to resolve.

Here's how I work through the tradeoff. To start, I accept that if some code needs to change to incorporate new features, there will be a regression risk whether we pay the debt or not. So, I fall back on my guiding heuristic: that we should pay the debt when doing so helps our productivity.

It's important to recognize and understand this risk or you might find yourself abandoning a rewrite and reverting back to the original, which was the case for Apple in 2015 when it reverted discoveryd back to mDNSResponder.[42] They have lived with this for 10 years so far. It's clear that the rewrite was unnecessary.

---

42 https://www.macrumors.com/2015/05/26/apple-discoveryd-replaced-with-mdnsresponder/

Abandoning a rewrite and reverting back to a working version is bad enough, but it could get even worse. In 2024, Sonos released rewrites of their mobile apps,[43] which broke a lot of working behavior. Unfortunately for them, they had changed so many aspects of their system that there was no way to go back, and they had to live with the broken version. Their CEO was fired in January 2025.

The risk of regression in a system is correlated to how many users depend on it. You can assess this with usage analytics, but it may be obvious if the code runs a core part of your offerings. If they are fundamental in your sales pitches, webinars, or highly promoted on your website, that's strong evidence that any regressions would be noticed and meaningful.

Regression captures something different from visibility, which is how meaningful a change to the codebase will be to your customers and stakeholders. The risk of regression is high, however, if they are already dependent on this part of the system before it is changed. Visibility is about how much users want the change; regression is about how much users already depend on what they have.

8. Uncertainty

Uncertainty is the risk that even a successful tech debt payment project won't produce its promised benefits. Fixing debt is supposed to make us more productive. New roadmap projects in this area should be easier to estimate, faster to implement, and more likely to be correct. If you are not sure that's what will happen, it should be represented in the uncertainty score. High values pull us back from doing this project.

---

43 https://techcrunch.com/2024/10/01/sonos-outlines-turnaround-plan-following-app-disaster/

Note that uncertainty is not the same as difficulty. Difficulty represents the risk that our estimate of a project's size is correct. Uncertainty assumes that the project is completed but fails to deliver results.

If you want to rate uncertainty low, you should have evidence that this type of project has been done before and works. Perhaps the debt is a known best practice or has been successfully deployed by others on your team or at your company. If you are inventing something from scratch, though, you should score uncertainty higher.

Score the Item Before the Kickoff Follow-Up Meeting

At the end of the kickoff meeting, we have ranked and grouped the items for each dimension. Here is the example I showed with misalignment:

Design System
FixMVC
Unit Tests
I18n
Zero Warnings
Dependencies
Folder Mess
Python3
Markdown
Permissions
Sockets
Sync Reliability
Window Memory

For purposes of this exercise, I assume my team and I are on the same page about the order and grouping, so the scores should descend from 5 to 0 down the list. To do that, I use the scoring guide (see Appendix 5) as a rubric, and these are the results:

Debt	Misalignment
Design System	5
FixMVC	5
Unit Tests	5
I18n	3
Zero Warnings	3
Dependencies	2
Folder Mess	2
Python3	1
Markdown	0
Permissions	0
Sockets	0
Sync Reliability	0
Window Memory	0

I want to make sure that I back up my scoring for the most misaligned items (in red and rated a 5), so I will note the places in our style guide where we have made agreements to work on this problem.

If someone at the meeting thinks, for example, that an item was mis-scored, then we could show them the evidence we found and see if they know of something we missed.

It is also possible that when I went to score the items, I found good reasons for why the list's order is wrong. Perhaps, after our meeting, our CTO announces a 1-year goal to get off Python 2.x across the organization. Now the Python3 project is much more misaligned than a "1" and needs to be moved up. Changing the order of the items that we agreed on requires justification.

The team scores the items for all eight dimensions and fills in a table displaying a column for each one (see the resources in Appendix 1 for a link to a template in Google Sheets).

Filled in, it looks like this:

Debt	Visibility	Benefit of Paying	Misalignment	Volatility	Cost of Staying	Resistance	Size	Cost of Paying	Difficulty	Uncertainty	Benefit of Staying	Regression
Dependencies	0	1.00	2	1	1.00	1	2	2.00	2	0	1.00	2
Design System	3	4.00	5	3	3.50	4	1	1.00	1	0	0.50	1
FixMVC	0	2.50	5	3	4.00	5	3	3.00	3	1	1.50	2
Folder Mess	0	1.00	2	5	4.50	4	1	1.00	1	1	1.00	1
I18n	5	4.00	3	5	2.50	0	5	4.00	3	0	0.50	1
Markdown	5	2.50	0	1	1.00	1	1	1.00	1	1	1.00	1
Permissions	3	1.50	0	1	2.50	4	2	3.00	4	3	3.50	4
Python3	0	0.50	1	4	3.50	3	3	2.00	1	0	1.00	2
Sockets	4	2.00	0	1	3.00	5	5	5.00	5	4	4.00	5
Sync Reliability	2	1.00	0	1	1.00	1	2	2.50	3	5	5.00	5
Unit Tests	0	2.50	5	5	5.00	5	5	3.00	1	0	0.00	0
Window Memory	5	2.50	0	2	1.00	0	3	2.00	1	1	1.00	1
Zero Warnings	0	1.50	3	4	2.50	1	3	2.00	1	0	0.50	1

The details of this spreadsheet are the focus of the next two chapters. We'll use it to set the priority for discussion and to drive our decisions about what to do about each item.

CHAPTER 25

Prioritize Your Tech Debt Items for Discussion

Now that your team has met, created a backlog, discussed the dimensions of each item, and scored them, it's time to start making plans for what to do. That's the focus of the kickoff follow-up meeting.

As a recap, let's quickly score the items on the **visibility** dimension.

The more a tech debt item's resolution would be noticed by customers and stakeholders, the more visible it is. Less visible debt would only be noticed internally, by engineering. With that in mind, I would rank and group them like this:

I18n
Markdown
Window Memory
Sockets
Design System
Permissions
Sync Reliability
Dependencies
FixMVC
Folder Mess
Python3
Unit Tests
Zero Warnings

I rated **Markdown** as one of the highest because at Trello nearly every customer uses features that require Markdown parsing, and even though it was mostly working as expected, it still generated a lot of high-priority support cases and bugs. *That's what a highly visible tech debt problem feels like.* Customers would get benefits and so would non-engineering teammates.

Conversely, the **Folder Mess** item made it harder to work on the code but fixing it would have (hopefully) had no effect at all on anyone outside of engineering. I couldn't explain it to a non-engineer. Even if I did, no PMM would mention it in a webinar and no salesperson would use it in their sales pitch. It's normal that items in our tech debt backlog have low visibility because if they were crucial to our customers, they would be in our regular roadmap, not here. It's also true that we shouldn't see items where the pay force is only driven by visibility, because then it's not really tech debt (it's just a regular bug or feature).

The next step is to score the items from 0-5. The expectation is that items in the same group will have the same score, and that the scores will descend. I end up with these scores:

Debt	Visibility
I18n	5
Markdown	5
Window Memory	5
Sockets	4
Design System	3
Permissions	3
Sync Reliability	2
Dependencies	0
FixMVC	0
Folder Mess	0
Python3	0
Unit Tests	0
Zero Warnings	0

I put scoring guides for all of the dimensions in Appendix 5. Before the kickoff follow-up meeting, use them to create a spreadsheet with all of the columns filled out. When you're done, you should have a table that looks like this:

Debt	Forces that indicate we should pay						Forces that indicate we should stay with debt					
	Visibility	Benefit of Paying	Misalignment	Volatility	Cost of Staying	Resistance	Size	Cost of Paying	Difficulty	Uncertainty	Benefit of Staying	Regression
Dependencies	0	1.00	2	1	1.00	1	2	2.00	2	0	1.00	2
Design System	3	4.00	5	3	3.50	4	1	1.00	1	0	0.50	1
FixMVC	0	2.50	5	3	4.00	5	3	3.00	3	1	1.50	2
Folder Mess	0	1.00	2	5	4.50	4	1	1.00	1	1	1.00	1
I18n	6	4.00	3	5	2.50	0	5	4.00	3	0	0.50	1
Markdown	5	2.50	0	1	1.00	1	1	1.00	1	1	1.00	1
Permissions	3	1.50	0	1	2.50	4	2	3.00	4	3	3.50	4
Python3	0	0.50	1	4	3.50	3	3	2.00	1	0	1.00	2
Sockets	4	2.00	0	1	3.00	5	5	5.00	5	4	4.50	5
Sync Reliability	2	1.00	0	1	1.00	1	2	2.50	3	5	5.00	5
Unit Tests	0	2.50	5	5	5.00	5	5	3.00	1	0	0.00	0
Window Memory	5	2.50	0	2	1.00	0	3	2.00	1	1	1.00	1
Zero Warnings	0	1.50	3	4	2.50	1	3	2.00	1	0	0.50	1

To help detect patterns, I highlighted any score over 4.0 in red. If you see a lot of red on the left side for an item, that means the forces to pay are high. The opposite is true if the red is mostly on the right.

Between each pair, I added a summary column. For example, between the visibility and misalignment, I added "Benefit of Paying," which is the average of the two. I did the same for each pair to show how it is either a benefit or cost of either paying or staying. The summaries will be used to find pay force and stay force as shown in this diagram:

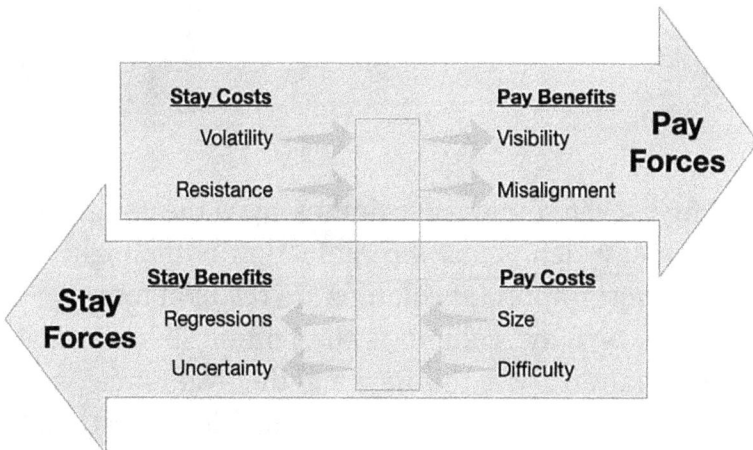

The first six numerical columns (after the item name) represent the top arrow pointing to the right. High visibility and misalignment scores imply there will be benefits of paying the debt (the first three columns), as do high volatility and resistance scores (the next three columns). High values in these columns act together as a force that indicates that we should do more about this debt.

But the next six columns (and the bottom arrow) are the dimensions that indicate the debt might not be worth acting on. They are what I call "Stay Forces."

The combination columns between each dimension will be used later to find the net force, which is useful in prioritization and getting a high-level overview. But when you are ready to discuss an individual piece of debt, it's better to use the eight individual dimension scores, as we'll see in the next chapter.

When you have a completed scoring table, you can use that to calculate a net force. I put that in the "Prioritize" tab of the spreadsheet (see Appendix 1: "Recommended Reading and Resources" for a link). Here's what it looks like:

Debt	Pay Benefit	Stay Cost	Pay Force	Pay Cost	Stay Benefit	Stay Force	Inverted Stay Force (for the chart)	Pay or Stay	Net Force	Joy	Net Force with Joy
Dependencies	1.00	1.00	0.72	2.00	1.00	1.14	3.86	Stay	-0.42	5.00	0.95
Design System	4.00	3.50	4.07	1.00	0.50	0.53	4.47	Pay	3.55	2.00	3.01
FixMVC	2.50	4.00	3.62	3.00	1.50	1.96	3.04	Pay	1.66	6.00	2.18
Folder Mess	1.00	4.50	3.04	1.00	1.00	0.72	4.28	Pay	2.32	5.00	3.14
I18n	4.00	2.50	3.62	4.00	0.50	1.96	3.04	Pay	1.66	3.00	1.57
Markdown	2.50	1.00	1.38	1.00	1.00	0.72	4.28	Pay	0.66	4.00	1.78
Permissions	1.50	2.50	1.65	3.00	3.50	3.62	1.38	Stay	-1.98	2.00	-1.98
Python3	0.50	3.50	1.65	2.00	1.00	1.14	3.86	Pay	0.50	3.00	0.95
Sockets	2.00	3.00	2.50	5.00	4.50	4.83	0.17	Stay	-2.33	5.00	-1.13
Sync Reliability	1.00	1.00	0.72	2.50	5.00	4.07	0.93	Stay	-3.35	4.00	-2.43
Unit Tests	2.50	5.00	4.07	3.00	0.00	1.14	3.86	Pay	2.93	2.00	2.38
Window Memory	2.50	1.00	1.38	2.00	1.00	1.14	3.86	Pay	0.24	1.00	0.00
Zero Warnings	1.50	2.50	1.65	2.00	0.50	0.93	4.07	Pay	0.72	1.00	0.37

The first seven numerical columns show the components of the net force, which is just a combination of the eight dimensions. The most important columns are the final four.

The "Pay or Stay" column is the bottom line on what you should generally do with this item and the "Net Force" is a score from -5 to +5 for the signed magnitude of this suggestion. Items to pay will have a positive net force, and items to stay (i.e., not pay) are negative. The gradient in the column is darkest for the highest numbers (the ones to discuss first). You should sort the data on this column, and then it will look like this (hiding the number columns for brevity):

Debt	Tr Pay or Stay	# Net Force	# Joy	Net Force with Joy	Bubble Size
Design System	Pay	3.55	2.00	3.01	2
Unit Tests	Pay	2.93	2.00	2.39	2
Folder Mess	Pay	2.32	5.00	3.14	5
FixMVC	Pay	1.66	5.00	2.18	5
I18n	Pay	1.66	3.00	1.57	3
Zero Warnings	Pay	0.72	1.00	0.37	1
Markdown	Pay	0.66	4.00	1.78	4
Python3	Pay	0.50	3.00	0.95	3
Window Memory	Pay	0.24	1.00	0.00	1
Dependencies	Stay	-0.42	5.00	0.95	5
Permissions	Stay	-1.98	2.00	-1.98	2
Sockets	Stay	-2.33	5.00	-1.13	5
Sync Reliability	Stay	-3.35	4.00	-2.43	4

After the "Net Force" column, I put a "Joy" column, which I mean for you to use to capture anything you think was not represented in the eight dimensions. My main use is to recognize that there is just some code I hate and would love to get rid of. You can see how this affects the priority by sorting on the "Net Force with Joy" column.

Plot the Forces to Visualize the List

Plotting each item on a chart may make it easier to visualize how they compare to each other. Here's a plot of the Pay Forces and Stay Forces. As you can see, debts with net higher Pay forces are closer to the upper right and colored light blue. Debts with net higher Stay forces are towards the bottom left and colored dark red.

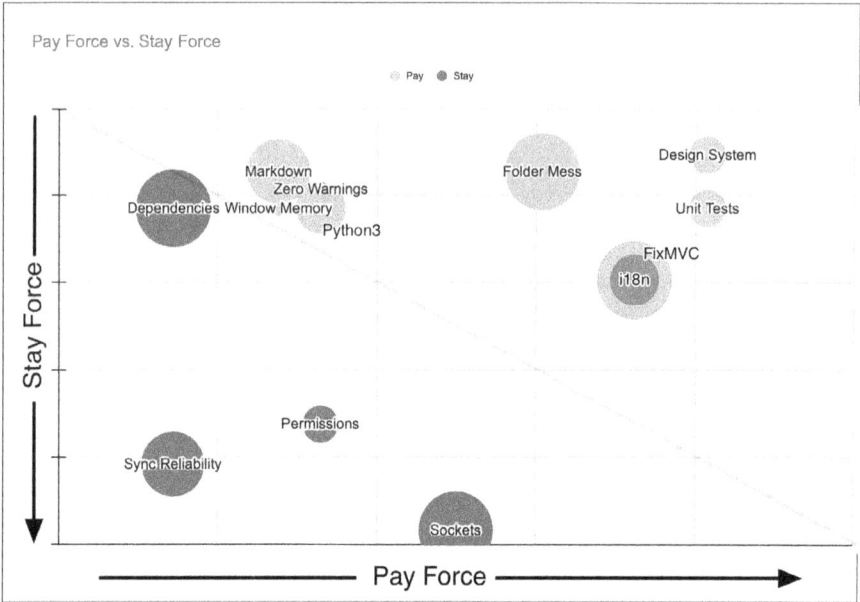

And to make sure you don't neglect "Joy," I use it to calculate each bubble's size. In the sheet, you could sort by the column that incorporates it into the net force, or you could imagine that large bubbles are higher than they appear and small bubbles are lower. In this case, it's enough to flip **Dependencies** from a stay to a pay and draws our attention to **Sync Reliability** and **Sockets**, which might warrant some attention (although not enough to totally pay it yet).

Again, use this plot as a guide to prioritizing your discussion, not as your final answer for what debt to pay. Your plan may differ based on more nuanced information. For example, if after seeing this plot you still thought that the team should discuss **Sync Reliability** first (after all, it has a high "Joy" score), it would be best to start by looking at the dimensions that are driving its current location firmly in the Stay bucket. In the next chapter, I'll go into more detail on how to make progress on debt like this.

At a glance, the **Design System** item has the most evidence supporting a plan to address it. This doesn't mean we won't do anything about the other issues. The various bubble positions and size will help guide our priority and decision to pay and how.

Use this visualization as a way to quickly compare the entire backlog. In the next chapter I'll show a radar chart that lets us visualize all of the individual scores of a few backlog items so that we can compare them. Then, we'll use that to decide what to do.

CHAPTER 26

Decide What to Do About Your Tech Debt

The actions we will take to deal with a tech debt item should depend on its scores across the eight dimensions of debt. When you look at a specific item of debt, you'll probably notice that certain dimensions dominate, and, accordingly, have the biggest impact on where it shows up on your bubble chart.

But as helpful as this chart can be, it's still just a way of getting a quick overview of your entire backlog. The next step is to pick a few items and compare the raw scores of the eight dimensions that impact them.

In this chapter we'll take a deeper look at the scores we recorded for each dimension, learning how we can use them. Along the way, I will make various suggestions (identified **in bold**) for how to best interpret your data and decide what to do. All of my suggestions are collected in Appendix 6.

To begin, let's go back and review the spreadsheet we created in the last chapter, but this time, I'll sort it by the "Net Force" column:

Debt	Forces that indicate we should pay						Forces that indicate we should stay with debt						From Prioritize Sheet
	Visibility	Benefit of Paying	Misalignment	Visibility	Cost of Staying	= Resistance	Size	Cost of Paying	Difficulty	Uncertainty	Benefit of Staying	Regression	Net Force
Design System	3	4.00	5	3	3.50	4	1	1.00	1	0	0.50	1	3.55
Unit Tests	0	2.50	5	5	5.00	5	2	3.00	1	0	0.00	0	2.93
Folder Mess	0	1.00	2	5	4.50	4	1	1.00	1	1	1.00	1	2.32
FixMVC	0	2.50	5	3	4.00	5	3	3.00	3	1	1.50	2	1.66
I18n	5	4.00	3	5	2.50	0	5	4.00	3	0	0.50	1	1.88
Zero Warnings	0	1.50	3	4	2.50	1	3	2.00	1	0	0.50	1	0.72
Markdown	6	2.50	0	1	1.00	1	1	1.00	1	1	1.00	1	0.66
Python3	0	0.50	1	4	3.50	3	3	2.00	1	0	1.00	2	0.50
Window Memory	5	2.50	0	1	1.00	0	3	2.00	1	1	1.00	1	0.24
Dependencies	0	1.00	2	1	1.00	1	2	2.00	2	0	1.00	2	-0.42
Permissions	3	1.50	0	1	2.50	4	2	3.00	4	3	3.50	4	-1.98
Sockets	4	2.00	0	1	3.00	5	3	3.00	5	4	4.50	5	-2.33
Sync Reliability	2	1.00	0	1	1.00	1	2	2.50	3	5	5.00	5	-3.35

Reading and comparing rows of data is hard, so after looking at this baseline I plot the values of each item on a radar chart. Let's take a deeper look at Markdown and Sockets. Here's a reminder of what they are:

Short Name ∨	Summary
Markdown	iOS markdown doesn't 100% match web
Sockets	Sockets system uses old tech and causes lots of bugs

And here's what it looks like when they are plotted against each other:

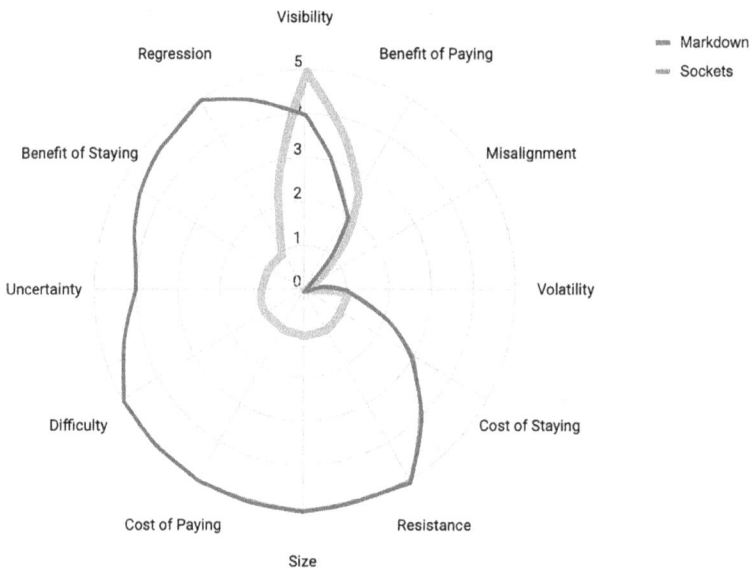

Starting at the top of the chart with Visibility and going clockwise, the right half displays the Pay forces. From Size on the bottom and going around clockwise, the left side displays the Stay forces.

Both items have relatively high visibility. In the case of the Markdown item, the benefit of paying is clearly higher than the cost of doing nothing, so it should probably be paid. But when visibility is the only high value, you should consider whether it should be classified as tech debt at all. If customers and stake-holders would like this item to be resolved, we could just roll the fix into our team's normal prioritization process that we use for deploying features and fixing bugs.

The situation is different for Sockets. Its debt has some visibility, but the forces indicating we shouldn't pay it are high, so it feels like making a payment might not be worth it. But if our team feels strongly that they want to make progress on this item, there are two strategies I suggest.

The first one is to confirm that the perceived visibility matches the actual one. If it's true that the tech debt in our sockets code is causing problems for customers, then there may be more support cases and reported bugs that are caused by this debt but aren't cataloged as such. If so, **use evidence to build a case that visibility is actually higher.**

This example comes from a real-life problem we had at Trello. Once we learned about a problem in our sockets code, we realized that a lot of weird things we had been seeing in support cases were being caused by it. We made a special tag and used it to mark cases that would be fixed by paying this debt. Doing this made us recognize that visibility was higher than we thought, and we got more support from our product manager to address it in some way. We had to fix the bugs, which we could do with a rewrite or by trying to address them with hacks in our current implementation.

If you are committed to fixing a system (by paying debt or working around it), you have to increase your volatility score because the codebase with the debt is going to change. For us, the net effect of all of this drove us to pay the debt with a rewrite that would result in a simpler sockets system. But your score changes might not be enough when the forces on the other side are too large.

In addition to increasing pay forces (on the right), it could also be true that the dimensions driving the stay force (on the left) can be reduced. You may have overestimated the difficulty dimension because you don't have a good plan, and your uncertainty is driven by your belief that your fix won't work. **Try to de-risk the plan by doing a timeboxed project spike.** Your goal is to reduce the debt's size and difficulty by making some initial payments and, as a result, getting a better estimate of how much effort it will truly take to pay the debt. You might also discover that your uncertainty about the benefit of the project was too high if the spike code works. Over time, the values on the left side of the radar chart would go down and help the pay forces overtake the stay forces.

Dealing With Less Visible Tech Debt

Both of those examples (Markdown and Sockets) had high visibility, so we knew that our work would be appreciated by our customers. But as you know, tech debt is often invisible outside of engineering, with impacts on developer productivity that go unnoticed by customers (not to mention the rest of the company). So now let's look at some examples of tech debt with low or nonexistent visibility: **Unit Tests**, **Folder Mess**, and **Zero Warnings**. Here's a reminder of what they referred to:

Short Name ⌄	Summary ⌄
Unit Tests	Our project has no unit tests
Folder Mess	The code folders don't match Xcode groups
Zero Warnings	There are too many compiler warnings and we can't see if there are real issues

And here's what it looks like on a radar chart:

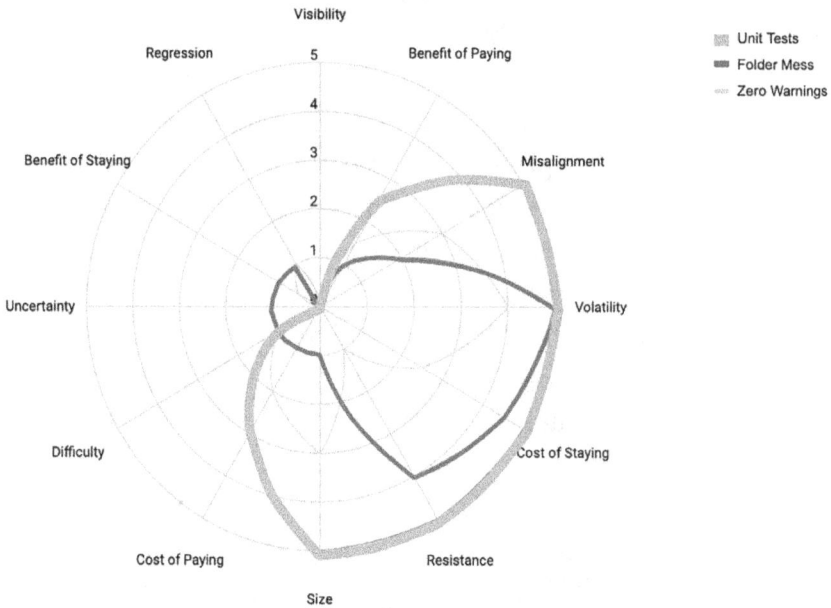

No one outside the team is going to care about **Folder Mess** (in dark red) because it's not going to affect them. On the other hand, this debt affects almost every line of code we try to write because it's hard to find the right file to do our work in. The good news is that it's not going to take a long time to fix (low size and difficulty), we know it will work (low uncertainty) and it's unlikely that doing so will cause any bugs (low chance of regressions). **Just fix it.**

Not having **Unit Tests** (shown in thick light blue) also affects productivity because it's risky to change any code, but adding tests will not be directly visible. It has a high misalignment score because it was against our established team value that new code should be tested. When we started, we had a three-year-old codebase without tests, and it was too much effort to add tests all at once. In this case (with low visibility and high cost to pay, but otherwise high paying forces), I recommend that you

enforce your values in reviews of new code. Even better, **automate adherence to your coding standards in CI.** Doing so will pay this debt over time and directs the payments at developer productivity that pays off immediately.

For **Unit Tests**, it's easy to gather a coverage metric. After you do this, **propose a goal and build a dashboard that shows your progress against it.** It's important that this dashboard is placed somewhere everyone will be able see it. When the Trello web team decided to replace CoffeeScript with TypeScript, they published a metric in Slack showing their progress on a regular basis. Doing this will add some visibility to the debt (and its reduction) because now it's possible to understand it without reading the codebase. For internal issues like unit tests and programming language updates, this dashboard helps explain the issue to engineering management all the way up the organization chart. It still won't matter to non-engineers or customers, so it will never get the highest visibility scores. When it's possible, though, it's worth some effort to help your manager show progress to their manager and all the way up to the CTO.

One thing we did on our team to make sure these dashboards weren't ignored was **allow dashboards that show visibility to create incidents** if they show a metric is out of bounds. We did this by pushing the data to our normal observability systems and building the dashboard the same way that DevOps did for the production systems.

Zero Warnings (a project that would fix all compiler warnings) had some effect on productivity since it made it harder to detect a truly important warning among all the noise of thousands of warnings. It's unlikely that this debt was visible, and, in this example the team had not documented a stance on warnings in its coding guide, so we couldn't justify doing much, so we decided to **ignore it**.

I have wanted this aspirational "zero warnings" project in every job I have had, but it was never the most important thing to do. If you have something like this and don't want to ignore it, you could get a higher misalignment score if you **propose that your team adopt a coding standard that addresses the debt over time.** If that works, then update your misalignment score and use the tactics directed at getting alignment (from the **Unit Tests** example above). If the team doesn't agree that this should be enforced in the guide, **be the change you seek: model the behavior you want in small doses** by fixing warnings in the code you update and in your spare time (or when you are given time to do what you want, see Chapter 14: "Remove Tech Debt When It's Your Decision"). One thing that's nice about this debt is that you could make progress in any random fifteen-minute time block, and it would probably make you happy. Others might join in.

Use Scores as a Way to Guide You

This was not an exhaustive list of examples. But hopefully they are a useful starting point on how to use analyses of dimensions to guide your tech-debt-paying plan. The goal is to make sure any plan is proportional to the problem it is attempting to solve and to try to make some progress each quarter, so that your overall debt level slowly reduces.

Use the scores as a guide to what to do but remember to listen to your intuition as well. If you think a score is wrong, then find additional evidence that either confirms or refutes your hunch. If a high score in one area is driving you to do something you feel is wrong, you can make progress by directing your actions to changing just that score. Here are some examples:

When you have:

- high uncertainty, propose a spike
- low misalignment, propose a new standard
- high size, chip away at it
- low visibility, make a dashboard to prove the debt causes customer problems

I cataloged the suggestions in this chapter (and many more) and how they map to the eight dimensions in Appendix 6, "Tech Debt Dimension-Driven Strategy Catalog."

Agree on a Plan and Publish It

Now that you've assessed each debt, you are ready to make a plan. Go back to your debt backlog and add a new column, "[Quarter #] Technical Debt Plan." Next to each item of debt, briefly explain any action you're going to take (or lack of action), how much time you will spend on it, and who will own it. Monitor progress and revisit this plan at the next recurring meeting.

Here's an example, taken from my sample backlog in the last chapter:

Short Name	Summary	Plan for Next Quarter	Owner	Budget (week)
Dependencies	Dependencies are updated haphazardly and get out of date	Update dependencies to latest now. Ongoing, do it right after each release to at least point releases.	LL	2
Design System	It isn't easy to match the design system	Implement a standardized font and color system.	LL	1
FixMVC	Our UI code isn't consistent or testable	Fix only if a screen has major changes. Add a guideline to our standards documentation.	Team (JD to Doc)	1
Folder Mess	The code folders don't match Xcode groups	Fix in the next sprint.	LL	1
I18n	The app is not internationalizable	Ignore until the product team decides to i18n the entire product.	N/A	0
Markdown	iOS markdown doesn't 100% match web	One week spike to see if porting is feasible. If so, create a plan and estimate for the following sprint.	LL	2
Permissions	API permissions are represented by guard-statements, not data	Propose to PM and remove from this list.	N/A	0
Python3	Update to Python 3	Ignore	N/A	0
Sockets	Sockets system uses old tech and causes lots of bugs	Tag any bug related to sockets. Three-week project spike this quarter. Create plan and estimate.	JD	5
Sync Reliability	We don't know what's causing our synchronization problems	Build an observability dashboard to show sync errors. Work with PM to define levels for Excellent, Acceptable, and Unacceptable and a plan of what to do at each level.	LL	2
Unit Tests	Our project has no unit tests	Add a unit-test target to our project and run on CI. Document a standard for new code.	LL	1
Window Memory	Our windowing system uses too much memory	Rewrite the memory usage code.	JD	6
Zero Warnings	There are too many compiler warnings and we can't see if there are real issues	Ignore	N/A	0
				21

There are a few things to note about this plan:

1. The bulk of the work is being done by JD and LL. In this fictional example, we are assigning two of our ten engineers to work on engineering-led items this quarter (see Chapter 20: "Make and Spend a Budget to Pay Back Tech Debt" for a refresher about the budget and allocation).
2. The total budget is 21 weeks. For a quarter with two people, we have a total of 26 weeks of engineering-led time, but we have other work besides tech debt.
3. The time budgeted is meant to be a hard cap, not an estimate. It's ok to need less time, but work has to stop when the cap is hit.
4. The highest priority items are going to be fully addressed, but they are small. Some items with high stay forces (e.g., Sockets) are going to be worked on to try to mitigate their risks first.
5. One of the items (e.g., FixMVC) is being done by the team in the natural course of their work with just the code they encounter. We are acknowledging that this new requirement has some overhead, but our intention is that in the future, this is baked into project estimates and has minimal impact because we're doing it to gain short-term productivity (see Part 2 for all of the ways this can happen).

At the end of the Tech Debt Kickoff Follow-up meeting, make a recurring quarterly meeting to go over the progress and manage this backlog.

The Recurring Tech Debt Meeting

Between recurring meetings, the team (mostly LL and JD) should be working on the plan. There are a few ways this spreadsheet will be updated until the next meeting.

1. JD and LL should add a summary update on the backlog items.
2. They should propose new scores when work is being done to change them. For example, if we are building a dashboard to increase visibility, they should propose a new visibility score.
3. The team should add new items.

A full sample agenda for recurring tech debt meetings is in Appendix 4, but generally they have the same goal as the follow-up did, which is to produce a plan for the next quarter. The main difference is in how they start.

Now that we have made some progress with tech debt, we'll start the meeting by going over it. Celebrate wins. When an item has no more development work left, I recommend that you make a plan to monitor its impact and communicate that to the rest of the company.

If completed items have high visibility, then make sure you demo that to your colleagues outside of engineering. Explain the customer impact (not just the internal impact) so that they can use that in marketing material and sales pitches.

Low visibility items are still meant to improve our own productivity. It's important to gather evidence of that effect and to communicate that to other engineering teams and up the engineering org chart. The next part of this book explores engineering leadership and their perspective on technical debt. It's meant to help teams learn how to influence them. A big part of that is helping them understand the effects of your work and how it relates to their goals. Let's see how.

PART FOUR
LEADERSHIP PRACTICES

I've been all over the org chart in my career. I've been the head of engineering for a startup with a dozen engineers; in charge of a few dozen engineers separated from me by a layer of managers; the principal engineer at a startup where I reported directly to the CTO; and a lead or manager in larger companies, several layers removed from the CTO.

Ironically, I learned the most about leadership when I was furthest removed from it. At Atlassian there were about four management layers between me and the CTO, who was my great-great-great-grandboss. As expected, the cadence of my interactions decreased with each layer. Still, there were things our CTO and middle managers did that helped me clear technical debt.

That's the focus of this part of the book. We'll look at ways that teams and leadership influence each other in large organizations. The fact is, you will be in a much better position to address tech debt if the CTO values it—even if they don't fully understand it. But they have goals too. The more you incorporate their needs and concerns, the more likely you are to get support.

We're trying to scale what happens naturally when you're small. CTOs at small companies still write code. They know everyone on the team. They get repeat exposure to your work and learn to trust you. You hear their priorities constantly and start to internalize them. They also feel the productivity gains you deliver. If that's your situation, the team practices outlined in Part 3 might be enough.

But at larger companies, where interactions between the CTO and engineers tend to be indirect and infrequent, their nature is different. In these organizations, communication needs to be clear and memorable—in both directions. It can't rely on being face-to-face. Your CTO isn't in the water with you anymore. They won't feel the resistance. They won't know when it's gone.

At a large company, your tech debt isn't always important enough for the CTO to get involved. Save your influence. If debt is a significant problem, though, these chapters will give you some strategies for getting the resources and time you need. You will learn how to align tradeoffs with the goals of the company. This mindset is useful whether you use it for tech debt or something else.

In my experience, to help teams pay debt consistent with a company's goals and strategy, engineering leaders must do four key things: (1) assess (and share) how bad your tech debt problem is; (2) set the constraints for addressing it; (3) give assurance that engineers have permission to clear debt within those constraints; and (4) tell engineers to communicate back the results of their debt-clearing work. We'll address all four in the next few chapters.

CHAPTER 27

Assess Your Tech Debt Problem

As a CTO, you might be far removed from the codebase. But in case you're wondering: yes, there is *definitely* tech debt in it. Things just happen.

I have been forced to deal with debt problems in a project's first few months. I once open-sourced a React UI library with some demo apps. This was right around the time that functional components were being introduced to React. Since they were new, I used class components instead. It was the right call at the moment, but it was inevitable that they would become debt when functional components became mainstream.

On a much larger scale, in the 90s I worked at a company that made a desktop application that ran on Windows and several flavors of Unix. The codebase was almost ten years old when I got to it, so there was tech debt for sure, but it was manageable. Then the web was invented, and the entire codebase became debt.

Small amounts of debt can be cleared by teams on their own without upper management intervention. Most of the tech debt I've paid down during my career was done as part of my regular work, as I described in Part 2. If the debt was too large to pay on my own, my teams used practices like the ones I described in Part 3. Most problems can be dealt with over time using those strategies.

Sometimes, though, tech debt issues become too big for team members to tackle on their own without more organizational support. An engineering team is under pressure to deliver on product requirements and typically doesn't have the authority to make the trade-off between delivering them and clearing tech debt. In the first three parts of this book, I stressed that individual contributors and teams should look for ways to clear tech debt so that it has immediate and visible results or (if the problem is completely internal) try to make the developer productivity benefits more evident. But these core practices might not be enough when the problem gets out of hand.

Of course, it would be great if you never get to this point. To make sure, monitor your tech debt problem. There's no better way to do this than talking with your engineers.

Talk about Tech Debt at Skip-level Meetings

At the companies where I worked, it was common for engineering managers higher in the org chart (the ones not directly managing engineers) to regularly meet with individual contributors. They'd often have a low-key meeting with the goal of building a relationship. These often took the form of group meetings where the leader was mostly listening or communicating broad principles.

These kinds of casual check-ins are the perfect time to get a feel for how tech debt is affecting your teams. Your goal during them should be to understand if you need to be involved and, if so, to what extent. Make tech debt a regular topic when you meet.

I believe that direct skip-level meetings between engineers and management are essential no matter how large a company's org chart is. If they feel infeasible for you right now, reduce their frequency, increase the group size, or limit attendance to the most senior engineers that still write code. But make sure that meetings happen.

I said "that still write code" because tech debt is a problem that directly affects engineers when they write code. In the same way that you might talk to users of your software to learn their pain points, you need to talk directly to engineers to learn theirs. You don't want a second-hand account from their manager.

Ask About Tech Debt in Surveys

Another way to engage with engineers is through surveys. If you already have some surveys in place, add a question or two about tech debt if there aren't any. Ask about the frustrations that result from reduced productivity, a major consequence of having too much debt. If you are not currently surveying at all, then add a short quarterly survey to monitor the health of the engineering team.

I frequently received surveys at Atlassian. They'd repeat the question, "Are you proud of the work you do here?" The answers were multiple-choice so that they could be aggregated. Each question also had a qualitative part where the respondents could explain why they answered the way they did. When I scored it low, having too much tech debt or not enough resources to deal with it were common reasons why.

Make Sure Your Strategy Is
Appropriate to Your Lifecycle

I had been working at Kofax for about a year when I got invited to our annual customer conference. My office was in Massachusetts and the conference was in San Diego, so I jumped at the chance to get out of winter for a week.

I didn't know it before I got there, but the keynote speaker was Geoffrey Moore, the author of one of the most important books about the tech business to ever be published, *Crossing the Chasm: Marketing and Selling Disruptive Products to Mainstream Customers*. In it, he described the transition from the very beginning of the customer adoption curve to the next phase as a "chasm." A chasm that you couldn't cross by just making incremental changes. You had to jump. A colleague gave me a copy in 1996, and I had thought about it frequently since then. I couldn't wait to hear what Moore had to say more than ten years later.

Before the conference, Moore had been spending time with our leadership, and they asked him to tell the conference attendees about the rest of the adoption curve, which he described in his 2005 book, *Dealing with Darwin: How Great Companies Innovate at Every Phase of Their Evolution*. Its core insight is that the product development strategy an organization chooses should depend on where the product falls on the adoption curve, from invention to its eventual decline.

The invention phase is well-studied. Moore's ideas are like others' who have written in this area. From Ben Horowitz's influential blog post about "Wartime vs. Peacetime strategies" to *The Lean Startup* by Eric Ries, there is no shortage of recommended strategies for beginning a new software product. Moore goes further, however, by offering solutions that apply to the entire customer adoption curve, not just the beginning.

As a result, his suggestions are especially suited to tackling tech debt, which is bigger in more established products. Time helped create it, but so did success. The role of engineering leadership, then, is to make sure their team understands the unique needs of their products at every stage of their lifecycle, and what the company's tolerance for debt might be, given a product's current position.

For example, at the very beginning of a product lifecycle, there should be a high tolerance for debt. You don't have product-market fit. The goal is to learn, and, as such, code will be thrown out constantly. This is not an excuse for writing crappy code. I'm working at a startup right now. I use tests and clear code to go faster, but I only do enough to help me in the short term. When it becomes clear a piece is code is valuable and going to stick around for a while, I go back and add more tests to stave off regressions.

When a product is maturing, and it's clear that we will have faster growth (or we are already experiencing it), we should anticipate big architecture changes. It might even be better to complete them now, while the migration problem is easier. If we have evidence that what we currently have will break (and by what date), then it's inevitable that you'll either fix it or suffer.

Closer to the end of a product lifecycle it becomes time to focus on operational excellence and cost reduction. There are fewer ways to generate more revenue. Tech debt projects aimed at making the product cheaper to run become important when we have to seek profit by lowering costs.

During the final stage, product decline, tech debt projects should focus on reversing complex but more scalable architecture choices, because our problems are now the opposite of scalability. Novel tech that we created years ago might now be available as open source. So, we could replace our code with parts that are maintained by others. We'd fix our debt by deleting code.

Make an Assessment

When you have a clear understanding of your position in the market and your tech debt problems, you are ready to decide if those problems are worth your attention or not. If they are large and widespread or have effects that interfere with your goals, then it's time to act.

If your engineering organization is large and multi-product, tech debt problems may be unevenly distributed. If the problems are varied across the organization, the best thing to do may be to focus your decision making and assessment on a single product. Learn. Get a win. Then, apply that to the other products.

If you decide that your teams need your support, look for ways to influence your front-line engineering managers and tech leads. I think the best way to do this is by establishing constraints and then letting engineering teams make their own decisions, which I will go into next.

CHAPTER 28

Express Constraints

Deciding on objectives may be hard but making sure everyone knows them isn't. It's mostly repetition. What I haven't seen done well is expressing constraints. When it was made clear what my team was supposed to do and how much time was okay to spend on it, we could usually come up with an appropriate plan that would meet their guidelines. Without that, you have to guess and keep asking for approval, which wastes time. It's another feedback loop to optimize.

When your team has an objective but no sense of the constraints, there's a risk they'll propose a fix that's disproportionate to the value of the objective. For the most important goals, it may be okay to propose spending months to achieve them. You might need bold proposals, and you are okay dealing with feedback loops to help shape them.

Having to deal with approval feedback loops can cause problems, however, when it comes to tasks that are important but further down the priority list, like managing tech debt. If you have to ask for permission every step of the way, it's an incentive to just give up. Fighting tech debt often involves a lot of small tasks that are done regularly in support of other work. It's much better for everyone if upper management defines an overall strategy and budget.

It Starts with Values

Most companies have a mission statement and a few articulated values. In my experience, these are largely not well-known throughout the organization and not used to make decisions. That's a missed opportunity.

What do they have to do with tech debt? A lot, actually. Struggling with tech debt is a major activity of an engineering team. Every engineering team on earth deals with it. So, if you are a tech company and your company values do not offer some clue as to how your engineers should approach their tech debt activities, there's a strong chance their daily decisions about it won't align with what you want. If you want alignment, they need to know what you value.

True, top-level company value statements are often too generic. At Atlassian, we had five company values; the one that seemed closest to touching on tech debt was probably "Build with Heart and Balance," but honestly, I didn't really understand what it meant because it was so abstract. The other values were easier to practice but didn't apply to tech debt.

At Trello I had a little more to work with. I could apply the statements "Don't do Nothing" and "Fix it Twice" to my tech debt, but even still, there were lots of ways to interpret these statements. To bolster our understanding, during Town Halls, our leadership would tell specific stories showing how our values were being applied. Regardless of the company I worked for, I always welcomed it when our CTO would explain how our company values should be applied to our engineering work.

Value statements that are pithy are easy to remember and use. Once explained, we at Trello knew what "Don't do nothing" meant. Meta's mantra to "Move fast and break things" is another good example (of pithiness). Every Meta employee can easily intuit the company's vibe and let it guide their decision-making.

I may not agree with where they are going, but I like how they are doing it together.

Similarly, when I was a teenager, there was a factory in Queens with a sign that said "Perfection is not an Accident" right under their name. I saw it every day on my commute to high school, and, decades later, I remember the value but not the company. I have no idea what they made, but I have a good guess on their approach.

Create a message for your team that's similarly pithy. Ideally it will provide some basic guidance to team members any time they are forced to weigh tradeoffs. Whatever that is, try to boil it down to something people can remember and refer to. Tell stories of the value in action. Put it on a T-shirt.

Your Budget Is a Statement of Your Values

Talk about values is important. But nothing communicates your values better than your budget. As a senator, Joe Biden once said, "Show me your budget, and I'll tell you what you value." If you value paying tech debt, you must express that with money somehow.

In Chapter 20 ("Make and Spend a Budget to Pay Back Tech Debt"), I recommended to teams that they split their budget in two. Some teams are inventing and spend more on product-led work. Some are managing legacy products and need to clean up more. The budget is how we put our values into action.

Here's a simplified example of how you might set up an engineering-led budget to use towards tech debt in an organization with three products (A, B, and C) and multiple teams. Let's say that overall, for the business, you think you could allocate 25 percent to engineering-led work. The total should add up to that, but it won't be evenly distributed among the products.

In this scenario, ProductA is a new concept that has just launched an MVP, so it's only built by one team. Its engineering allocation would accordingly be low, say, 10 percent of its budget. They don't have debt problems, but there's always some engineering-led work to do.

ProductB is a cash cow, but it's no longer growing. It has a couple of teams working on it because its customer base is still large and important. This product pays a lot of the bills. The work you do on its code may be focused primarily on reducing its deployment and maintenance costs. You have very few ideas for new features, and the product owner is mostly prioritizing bugs and support cases. Because it is in the part of the lifecycle where you concentrate on cost reduction, the engineering-led allocation might be 40 percent of the total.

ProductC reflects the future of the company and is experiencing steady growth. The codebase is several years old and has a reasonable amount of tech debt. Maybe ProductC is big enough that it needs to be broken down further. Perhaps some teams use 15 percent of their budget for engineering-led activities, and some use more.

This table summarizes the spread across all of the products:

Team	Team Size	Engineering-Led
ProductA	10	10%
ProductB	40	40%
ProductC: Platform	10	20%
ProductC: Apps	40	15%
Total	100	25%

As you can see, the total engineering-led allocation is 25 percent across all products. As part of my calculations, I considered the size of the team working on every product. So, ProductA has one engineer doing engineering-led work (10 percent of 10). ProductB has sixteen engineers doing it, and ProductC has eight total (two plus six). The total number of engineers doing this work is twenty-five, which matches our target allocation of 25 percent. If you have a concrete allocation in mind, play around with the numbers and percentages to arrive at your target.

These numbers are just an example. The takeaway is that you should have an overall target and then allocate it to individual products or teams based on their situation. This can be done from the top down, or you can set the overall goal but push the details down to your product team managers to negotiate. Then check that it rolls up to something you approve.

When I was an individual contributor in a large organization, having a priority be expressed through a budget helped me understand the company's goal much more clearly, whether the budget was large or small and whether I agreed or not. When the goals were vague, it was harder for my team to make decisions because we couldn't reconcile different opinions about what the goal meant. It's trite to say that "money talks," but that actually works here.

OKRs

Your other tool for communicating values is your Objective and Key Results (OKRs)—or equivalent. If tech debt is a major concern, it should be expressed as an OKR. Your senior engineers are more likely to follow it. More importantly, they can use it to persuade others.

It would be rare for tech debt to rise to a high-level OKR at a large organization. Generally speaking, they are meant to articulate a small number of objectives, maybe even one. In recent memory, the big exceptions have involved major platform shifts, like Facebook going "mobile-first," Bill Gates's internet memo, and nearly the entire tech industry moving desktop and server-based solutions to the cloud. Most of the time, your OKRs won't deal with debt.

However, further down the org chart, there might be teams with large tech debt issues. In the final chapter of this part, I describe a time when I worked on a tech debt project for almost two years because it was the most important activity of my team. However, that team was just five people in an engineering organization of hundreds. The rewrite was our only OKR, and showed up in the OKR of our division, but our work was only a small percent of the total budget, so it wasn't the main thing on the mind of our CTO. In a small organization, a project of this size would be all that most people would think about.

The part of the OKR that drives activity is the Key Result, which is how we know we have achieved the Objective. Throughout this book, I have stressed that tech debt projects should produce a positive result that is visible to the rest of the organization, and defining the Key Result is a way to set up the project to do that. If your debt isn't visible in a final product, then find another way to showcase its success. A dashboard of internal indicators showing if the debt project succeeded is a common option.

Promotions and Recognition

Incentives matter. In the end, your employees will incorporate your values when it comes to tech debt if their performance reviews and promotions depend on it. So, if you recognize engineers publicly for their tech-debt work (at a major anniversary, for example), or if an engineer receives a promotion because of their tech debt work, everyone will come to believe what you say about tech debt isn't just words.

When I mentor developers about job interviews, I advise them to ask prospective employers about employees who were recently promoted. What did the engineer do to merit the promotion? Then, when the mentee joins a company, I tell them to look at the job description for engineers at their level and the level just above theirs, so they know what the company values. I caution them not to depend on what the company says, but to look at what they do—especially when they have to make a hard choice between jobs.

Another thing I look for is how I'll be managed. It's easy to say that you don't micromanage, but that's what will happen if you don't communicate constraints. Bad communication leads to proposals that get rejected. It leads to projects that don't deliver what you want. Worse, it makes you think you need to constantly ask what's going on. The next chapter offers a solution to that.

CHAPTER 29

Give Autonomy and Require Accountability

Your engineering teams know the best ways to address tech debt. They feel the pain. They can tell if it's working. So, after you establish some boundaries, you have to let them go ahead and do it.

Sometimes, though, a debt problem is big enough to require your involvement. If this happens, you need to develop a system for staying informed on how things are going. Here are some techniques for doing this.

Ask for Observability

In Chapter 9 ("Make the Effects of Tech Debt More Visible"), I described ways that engineers can make code-level problems more evident to stakeholders outside engineering. The first story I shared explained how, at Atlassian, our team created a dashboard to show the reliability of our data synchronization system. I made tech debt easier to clear because stakeholders could see our progress. By making our debt (and fixes) more visible, we justified its cost and made it safer to change.

In Chapter 21 ("Add Tech Debt Rules to Your Style Guide"), I listed automated tools you could use to measure and report on style guide rules. Importantly, any of these tools can be used to set up dashboards that display problems directly related to the code.

There should be some way that engineers can show the effect of their tech debt work, even if it isn't evident by using the product. At Atlassian, my skip-level manager didn't tell us which metrics to track, but he us wanted to pick *some* measurement and use it to inform our decisions.

There's a downside to watch for. When faced with a mandate from management to reach a certain goal, savvy employees often quote Goodhart's Law, which states, "When a measure becomes a target, it ceases to be a good measure." To their credit, our management was able to lessen the dilemma somewhat by letting us choose what to track and how to fix it. They (rightly, in my opinion) understood that we were less likely to game a system we set up. For example, when a colleague set up a dashboard that displayed build times, we wouldn't have gamed it even if we could because the whole point was to make our lives better. We would just be hurting ourselves, since the dashboard we created was directly related to the pain we felt.

Track Budget vs. Actual

In the last chapter I recommended that you create a budget for each engineering team to use as they see fit for addressing tech debt. That's not the end of your responsibility, though. As CTO you need to make sure that that budget is not exceeded—but also that it's spent. Since a budget is just a way to allocate time, under-allocating time and money to tech-debt tasks means that there won't be enough time to pay it off. The budget is supposed to be a mandate. It's no use having a plan to address technical debt and developer productivity if it's always abandoned in the face of a looming product deadline.

Whatever budgeting process you use is probably fine. My main suggestion is to make it easy to compare the budget and actual spending. In Chapter 20 ("Make and Spend a Budget to

Pay Back Tech Debt"), I argued that the best way to spend an engineering-led budget line was by assigning individual developers to projects as opposed to trying to tag and track individual stories.

If a single person is assigned to work for a quarter on a tech-debt project, then we can assign all of their salary to engineering-led work. If you have a team of ten engineers, and the goal is to spend 20 percent on engineering-led work, then they just need to assign two engineers to work on the plan they built. This is enough to know that the budget is being spent.

Go Full Circle with Surveys and Skip-levels

In Chapter 27, ("Assess Your Tech Debt Problem"), I said that you should appraise where you are with your tech debt by using surveys. Your use of them doesn't need to be limited to the assessment phase—they can also help see if it's working. Use them to assess your code's health and better understand its effect on your developers. Since your team decided to work on a debt, how have their original survey answers changed, for better or for worse?

Additionally, if you hear directly in skip-levels that things are improving, ask for success stories that you can share more widely and share the stories that you heard in other meetings. Look for stories that improved customer satisfaction and/or the company's business. These meetings can spark a virtuous cycle where debt problems are identified, solved, and then celebrated. When a new debt emerges and the cycle restarts, they can be used as guides or cautionary tales.

The final chapter of this part is a story of the most ambitious tech debt project I was a part of and the lessons I learned.

CHAPTER 30

Give Big Rewrites
Enough Support
(or Don't Do Them)

I generally avoid large scale rewrites of code. During my career I have seen them fail or drag on much more often than I have seen them succeed. It's so hard to deliver any value from this work until it's complete, so it requires a lot of patience and commitment to keep it going.

But around twenty years ago, I participated in a nearly two-year long rewrite that did work. It was an enlightening experience, because it gave me important insights: a) what rewriting best practices look like and b) why it's so hard to apply them to most rewriting projects. Hofstadter's Law applies here: "It always takes longer than you expect, even when you take into account Hofstadter's Law."

Whenever my clients insist that they want to take on a rewrite, I share this experience to help them better understand what they're getting themselves into. Let's go over it.

In 2004, I was hired by ISO-NE, the nonprofit that manages the electrical grid in New England. Beginning in the late 90s, ISOs (Independent System Operators) were established by local energy utilities across the country to provide reliable service to residents. One of their most important tasks is managing the electricity market, which operates 24/7 and determines the prices the utilities are paid for power generation. These prices are shared with the public in the form of an online market publishing system.

When I got to ISO-NE, their market publishing system was a pile of Bash, Perl, PHP, and C. These scripts mixed database access, HTML generation, and logic in unexpected ways. Sometimes a C program would generate a PHP page in a cron job. It was a mess.

I was hired to help rewrite the entire thing as a clean Java-based system using (at the time) modern enterprise architecture patterns. I had been offered the job because I had experience with both the legacy languages and J2EE. Much to their credit, the organization seemed to understand the scope of the problem. They had doubled the engineering team size from two to four, adding me and a developer with a lot of Java Server Pages experience. Not long after, the team doubled again to eight, with contractors that would only work on the new system. They were even serious enough to break us out of a larger team and hire a manager who had done this kind of project before. They set a target date of nearly two years out.

Frankly, the size of the group and the timeframe seemed like overkill to me. But we got those resources because the situation was untenable; the code was fragile and our mission critical. Ultimately, we needed all eight of us and the whole two years to complete the task. I had been wrong.

We were given this budget because the power companies had made sure that the executives of the ISO didn't ignore things like the market publishing system. The board made sure we'd act by establishing several metrics measuring reliability. If any was not hit, they would set the bonus of every employee to zero.

They cared most about blackouts, and rightfully so. In 2003, the team that was monitoring the grid (some of whom were ex-Navy SEALs) had stopped the Northeast blackout from affecting most of New England because of their extensive training and quick reactions. The market publishing system wasn't *as* important, but there was little leeway for outages. Our code made our management nervous.

Given an appropriate budget, they trusted us to do the rest. The next thing we did was to make improvements to the legacy system along the way. Though I was hired to work full-time on the new Java code, after getting started we realized the best thing for me to do was to spend some time beefing up the legacy code, which would still have to run for the two years before being replaced. Because I was involved in both codebases, whenever I made a change to the legacy code, I either added it to the new code or updated the spec. By the end of the project, all of the full-time engineers were working on both codebases at about the same ratio. The systems were built in parallel, and both were kept running during an overlap period.

Our next good decision was kind of lucky, but it can be replicated.

By chance, our team also maintained a public document repository, which was implemented as a bunch of Perl and PHP scripts but was going to be migrated to an enterprise-level content management system at the same time.

We could have created templates that matched our current look, but instead we created a fresh design to make it more obvious to everyone that something was different. We branded both of these projects—the content migration and publishing system codebase rewrite—as Web Application Redesign Project, or Warp. All of Warp was released on the same day. In our launch, we mostly talked about the things people (non-engineers) could see: the new design, the easier way to update content, the content workflow, etc.

After the system had been up and running for a while, we could finally talk about how we had reduced the number of incidents and improved reliability metrics. Our customers, the electricity utility companies, really cared about this, but we waited until we had proof, not promises.

If the current state of your system is untenable and a rewrite can't be done piecemeal, you may not have a choice. A major rewrite may be the only option.

Here's what I learned. It's essential to make sure everyone in the company, all the way up to the executives, understands what needs to be done and agrees that it should be done. These projects are big and often fail when they overrun a low-balled estimate and get canceled. Resolving this upfront is worth the fight. This is arguably the most important lesson of rewrites, and one I saw again at Atlassian, fifteen years after the ISO project, when they executed a multi-year rewrite that transformed them from a mostly on-premises company to a cloud-based one.

Executing a large-scale rewrite will likely require you to bring on new developers, whether that involves hiring contractors or temporarily reassigning engineers from other areas of the company (if you can). To help this new group succeed, consider bringing in front-line managers who have done this before.

Finally, it really helped my ISO team to couple our rewrite project with a fresh design that was more visible to stakeholders. If you can find a way to combine a big tech debt project with a more visible one, you'll have something to show when you're done, and you can follow-up with metrics that display the benefits of the tech debt part after you have measured them.

PART FIVE
TOPIC-BASED CHAPTER GUIDES

I wrote this book with the expectation that you would read it from beginning to end, only skipping parts that felt irrelevant to your debt situation, if any. I began by articulating my overall philosophy about tech debt before describing debt-resolving techniques that applied at an increasingly bigger scope.

To recap, Part 1 outlined my assumptions and beliefs about debt and the value of paying it down. Part 2 focused on ways you can pay tech debt as an individual. Part 3 described how teams should work together to implement larger, system-wide tech debt fixes. And Part 4 covered how leadership at larger tech companies can help their teams succeed and how teams should communicate to them.

But you may have a problem with solutions strewn all over the book. In this final part, which is organized by topic, I highlight the locations in the book where you can find solutions to common debt-related problems.

CHAPTER 31

How to Increase the Amount of Time You Are in Flow

According to the DevEx model of developer productivity, the amount of time a developer spends in flow is one of the three leading indicators of productivity. Review the following chapters to learn more about getting into flow and the benefits of doing so.

From Part 1: Rethinking Tech Debt

- Chapter 4 ("Consider the Total Time Saved") introduces DevEx and its three components (including extending flow).
- Chapter 6 ("Pay Off Tech Debt to Make You Happy") connects flow to job satisfaction.

From Part 2: Personal Practices

- Chapter 8 ("Start with Tech Debt") is about how to get into the flow.
- Chapter 17 ("Borrow with a Short Grace Period") offers a technique for staying in the flow when you come across other distracting tech debt.
- Chapter 13 ("Make Progress While You Wait") shows you how to stay in the flow when you can't move on with what

you're doing because you are waiting for someone else (typically a code review).

- Chapter 19 ("End with Broken Code") explains how to preserve the flow you have at the end of a coding session and use it to start quickly at the next session.

CHAPTER 32

How to Shorten Code Review Loops

The most common feedback loop for developers is code reviews, and there are several ways paying tech debt can shorten it.

From Part 1: Rethinking Tech Debt

- Chapter 4 ("Consider the Total Time Saved") introduces DevEx and its three components (including shortening feedback loops).

From Part 2: Personal Practices

- Chapter 13 ("Make Progress While You Wait") gives you ideas for what to do while waiting for a code review.

From Part 3: Team Practices

- Chapter 21 ("Add Tech Debt Rules to Your Style Guide") shows you how to avoid excessive discussions about tech-debt resolution in a code review by making and documenting decisions about it beforehand.

I have also written extensively on code reviews on my blog.[44] My posts go beyond discussing tech debt and also offer suggestions for shortening code review idle times and reducing the likelihood that reviews will find problems.

---

44 https://loufranco.com/blog/category/code-reviews

CHAPTER 33

How to Shorten QA Loops

Along with code reviews (see the previous chapter), being forced to reopen items by QA because of new bugs or regressions is another common source of delays.

From Part 1: Rethinking Tech Debt

- Chapter 4 ("Consider the Total Time Saved") introduces DevEx and its three components (including shortening feedback loops).
- Chapter 5 ("Aim for Substitution") describes how to apply Liskov's behavioral substitution principles to tech debt payments to avoid regressions.

From Part 2: Personal Practices

- Chapter 10 ("Isolate Bugs into Small, Tested Functions") describes a way to ensure that you have really fixed a bug.
- Chapter 11 ("Fix Mistakes Twice") points out that a misunderstanding that results in a bug might show up in the codebase again and again, so your best option is to search for other instances on your own instead of waiting for QA to find them for you.
- Chapter 18 ("Turn Documents into Code") explains how out-of-date specifications slow down QA and are thus a type of tech debt.

- Chapter 19 ("End with Broken Code") introduces the concept of mutation testing, which can help you find undertested code.

CHAPTER 34

How to Reduce Cognitive Load

The amount of cognitive load you need to have to understand your codebase is another one of DevEx's three leading indicators of productivity.

From Part 1: Rethinking Tech Debt

- Chapter 4 ("Consider the Total Time Saved") introduces DevEx and its three components of productivity (including lowering cognitive load).

From Part 2: Personal Practices

- Chapter 8 ("Start with Tech Debt") shows you how to represent your learnings as new commits to decrease your cognitive load.
- Chapter 16 ("Fix Tech Debt as Part of Onboarding") shows you how to do small tech debt projects to quickly learn new development processes and tools.

CHAPTER 35

How to Reduce the Cost or Risk of Paying Tech Debt

From Part 1: Rethinking Tech Debt

- Chapter 5 ("Aim for Substitution") describes various ways to approach paying tech debt in a way that avoids regressions.

From Part 2: Personal Practices

- Chapter 12 ("Eliminate Debt to Make Estimates More Accurate") asks you to pay tech debt instead of adding it to an estimate.
- Chapter 14 ("Remove Tech Debt When It's Your Decision") shows ways to incorporate reducing tech debt into your regular workflow and how to leverage special events to pay down debt.

From Part 3: Team Practices

- Chapter 24 ("Score Tech Debt Along Cost-Benefit Dimensions") analyzes the costs of paying tech debt and the benefits of ignoring it. Review the sections on the **Size**, **Difficulty**, **Regressions**, and **Uncertainty** dimensions. High values in these dimensions indicate that a debt should not be paid. If despite this you still want to reduce the debt, a good first step is to focus on work that will result in lower dimension values.

- Chapter 26 ("Decide What to Do about Your Tech Debt") offers suggestions for reducing the size and risk of tech debt projects.
- Appendix 6 ("Tech Debt Dimension-Driven Strategy Catalog") builds on Chapter 26, offering additional strategies for reducing the size and risk of tech debt projects.

CHAPTER 36

How to Get Support to Pay Tech Debt

Some tech debt is too large or too hard to pay as you are working, so you will need additional support to resolve it. This is easiest if your debt is holding back development on a product, especially in a way that's visible to the rest of the business (i.e., non-engineering staff).

From Part 1: Rethinking Tech Debt

- Chapter 3 ("Couple Tech Debt Fixes with Value Delivery") argues that when discussing debt with stakeholders outside engineering, you should focus on what they will get from the project, not the debt payments themselves.

From Part 2: Personal Practices

- Chapter 9 ("Make the Effects of Tech Debt More Visible") shows you how to make internal impacts of debt payments more visible to your management.

From Part 3: Team Practices

- Chapter 20 ("Make and Spend a Budget to Pay Back Tech Debt") advises managers to create a dedicated tech debt budget and allow engineers to decide what to do with it.

From Part 4: Leadership Practices

- Chapter 28 ("Express Constraints") tackles approval loops that result in rejection when your teams don't know what the CTO wants.
- Chapter 29 ("Give Autonomy and Require Accountability") goes further on avoiding approval loops and micromanagement.

Part 4 is all about how leadership can help tech debt projects succeed. If you're looking for support from your own leaders, read these chapters to better understand their point of view.

APPENDICES

Recommended Reading and Resources

Below is a list of online resources from my personal website and a list of books and online content focusing on tech debt that has an outlook similar to mine. They cover a wide range of debt-related problems, so there's a good chance you will find at least one of them useful.

Online resources on my website

Resources for *Swimming in Tech Debt*:

This is where you'll find the spreadsheet that was the basis of Part 3 and links to my other related work.
https://loufranco.com/tech-debt-book/tech-debt-resources

As mentioned in Chapter 7, I will be maintaining an AI based addendum here:
https://loufranco.com/tech-debt-book/ai

If you want to know when I've updated anything, sign up here:
https://loufranco.com/tech-debt-book

Books

Accelerate: Building and Scaling High Performing Technology Organizations by Dr. Nicole Forsgren, et al.

Dr. Forsgren is a lead author on this book and many papers on developer productivity (discussed in Chapter 4). This book covers metrics related to deployments (e.g. DORA) that are good ways to see the ultimate effect of tech debt work (faster iterations and fewer failed deployments).

Working Effectively with Legacy Code by Michael Feathers

Describes the concept of a "seam," a place in the code where behavior can be changed externally. Seams are used to install mocks that allow you to isolate the code under test. Feathers outlines strategies to introduce them into systems with few tests or no test infrastructure.

Refactoring: Improving the Design of Existing Code by Martin Fowler

A catalog of safe, mechanical refactorings with step-by-step instructions. Following these steps will help you remove technical debt with a low risk of introducing regressions.

Tidy First?: A Personal Exercise in Empirical Software Design by Kent Beck

Offers a more recent take on Beck's philosophy of making constant small improvements to your codebase. I still love *Extreme Programming Explained*, but it's now over twenty-five years old and may feel a little dated (he was writing a SmallTalk system when it was published).

The Four Disciplines of Execution (4DX): Achieving Your Wildly Important Goals by Chris McChesney, Sean Covey, and Jim Huling

This book's recommendation to concentrate on leading indicators to meet goals has greatly influenced my thinking about tech debt, especially when it comes to thinking about its short-term productivity gains. It convinced me that DevEx metrics are good examples of leading indicators and should be a central part of any tech-debt reduction plan.

Flow: The Psychology of Optimal Experience by Mihaly Csikszentmihalyi

The seminal work on the concept of flow. Since developer flow is such a pillar of developer productivity, I recommend you read it.

Thinking in Systems by Donella H. Meadows

This book will help you understand how to model and influence large systems. My main takeaway was the importance of feedback loops, which are everywhere in software development processes.

A Philosophy of Software Design by John Ousterhout

This book's core message is about taming complexity in software systems, which is a big source of cognitive load. He introduces the terms *tactical programming* and *strategic programming* as two ways to work with complex code. He argues that the tactical approach to focusing on the change and not fixing the complexity is short-sighted and will not scale. Instead, he recommends a strategic approach which combines overall system simplification with getting the work done.

Online Resources

Quantified Tasks by Jason C. McDonald
McDonald's approach to task estimation, explained on his site, aligns very closely with my approach to tech debt scoring. https://www.quantifiedtasks.org

The Pragmatic Engineer by Gergely Orosz
In 2024, an excerpt from *Swimming in Tech Debt* appeared in this popular engineering newsletter. If you subscribe, you will get access to the archives where you can find articles on thriving as a founding engineer, migrations done right, and stacked diffs. Each of these articles touches on some aspect of technical debt. https://newsletter.pragmaticengineer.com

Can We Please Stop Talking About Tech Debt? by Emily Rosengren
She also thinks the debt metaphor doesn't work anymore. See why in this video:
https://youtu.be/DvfMOJaIzhY?si=LCCxmQ9ZQRXUSG-8

A Triple Bottom-line Typology of Technical Debt: Supporting Decision-Making in Cross-Functional Teams by Mark Greville, Paidi O'Raghalliagh, and Stephen McCarthy
This research paper categorizes tech debt similarly to the way I do in Chapter 24 ("Score Tech Debt Along Cost-Benefit Dimensions"). The authors' methodology, which involves surveying a corpus of published papers about tech debt to find patterns, is most useful for finding ways to avoid tech debt, whereas my taxonomy is meant to be used to apply to fixing already existing tech debt.
https://www.researchgate.net/publication/357738969_A_
Triple_Bottomline_Typology_of_Technical_Debt_Supporting_
Decision-_Making_in_Cross-Functional_Teams

Prioritizing Technical Debt as If Time & Money Matters by Adam Tornhill

This helpful YouTube video created by programmer and writer Adam Tornhill details metrics you can use to measure tech debt in a codebase.

https://www.youtube.com/watch?v=w9YhmMPLQ4U

What is Technical Debt? by Laura Tacho

I agree with Laura's opening that developers all disagree on what debt is. *Swimming in Tech Debt* represents my take. Laura's definition is similar (emphasizing the "negative impact of friction"). Laura is also an expert on developer metrics. I took her SPACE course and recommend her work highly.

https://lauratacho.com/blog/what-is-technical-debt

Tech Debt Kickoff
Meeting Sample Agenda

This meeting and its goals are described in Chapters 22 ("Schedule Regular Tech Debt Meetings") and 23 ("Rank Debt Items at the Kickoff Meeting"). Detailed descriptions of cost-benefit dimensions are described in Chapter 24 ("Score Tech Debt Along Cost-Benefit Dimensions").

Purpose

To create a backlog of tech debt items and stack-rank each against eight dimensions of cost and benefits.

Asynchronous Pre-work

1. Start a backlog document and instruct everyone to add items of technical debt to it. For each item they should provide a short name, summary, and a link to a longer description if needed.
2. Encourage team members to comment on and discuss the items before your meeting. This can be done in the document itself so all team members can view the comments.
3. Combine any duplicate list items.
4. If you end up with more technical debt than can be discussed in an hour, lightly prune and order the document

so that the most obviously important items are at the top and minor ones are moved to the bottom.

5. All participants should be familiar with the content of Chapter 24, which describes the eight dimensions of technical debt, which will be used at the meeting.

Meeting Agenda

Total time: 1 hour

Part 1: Complete the backlog (10-15 minutes)
The goal of this part is to come to a common understanding of the items in your backlog. If everyone has done their pre-work, this should go quickly. Use this time for making small clarifications.

1. Go through each item one by one, with the author of each item reading its name and summary.
2. After each item is mentioned, ask if there are any questions about it. Be careful not to get derailed. I recommend time-boxing a few minutes at most for each item.
3. Once you get through the full list, ask if there are any other items to add.
4. At the 15-minute mark, narrow the discussion to the most important items that are left.
5. Make sure everyone knows this is a living document that you'll return to in each meeting and that they can add to it at any time.

Part 2: Stack-rank the debt along various dimensions (30-40 minutes)
As a team, stack-rank the tech debt items on your list against each of the eight dimensions of tech debt. Here's an example

from Chapter 23 showing how items from a sample backlog rank across the Misalignment dimension:

Design System
FixMVC
Unit Tests
I18n
Zero Warnings
Dependencies
Folder Mess
Python3
Markdown
Permissions
Sockets
Sync Reliability
Window Memory

The list is in descending order. Design System item has the highest Misalignment score, while Window Memory has the lowest, i.e., Design System is the most misaligned with our values while Window Memory is the least.

Repeat this process for all of the other dimensions.

Part 3: Assign follow-up work for the next meeting (5-10 minutes)
In this meeting, the goal was to figure out where there is wide agreement and if we might need to gather evidence to resolve

disputes over rankings. If your group broadly agrees on the rankings along each dimension, scoring should be simple. If there is disagreement, it should be resolved before the next meeting.

To do this, assign the dimension columns (e.g. visibility, misalignment, etc.) in the spreadsheet evenly across your team to be scored. They should use the scoring guide in Appendix 5 and the detailed descriptions in Chapter 24.

Tech Debt Kickoff Follow-up Meeting Sample Agenda

Note that this meeting and its pre-work are explained in detail in Chapters 24-26.

Purpose

To create a prioritized tech debt backlog, make plans for addressing each item, and add new items in your main work-tracking system as necessary.

Asynchronous Pre-work

At the end of the previous meeting (the Kickoff Meeting described in Appendix 2), all of your tech debt items should have been stack-ranked across each cost-benefit dimension.

So for every dimension, there should be a list of each item in descending order of how much that item embodies the dimension (i.e., the most visible item should be at the top of the Visibility list, the most misaligned should be at the top of the Misalignment list, etc.).

Here's the example from Chapter 23 again, taken from the ranking against the Misalignment dimension. The ranking indicates that Design System, the list's highest-ranked entry, has the greatest misalignment among all items.

Design System
FixMVC
Unit Tests
I18n
Zero Warnings
Dependencies
Folder Mess
Python3
Markdown
Permissions
Sockets
Sync Reliability
Window Memory

Between the Kickoff Meeting and Kickoff Follow-up Meeting, team members should:

1. Meet or asynchronously score each piece of tech debt against their assigned dimension. Use the scoring guide in Appendix 5 to complete this task.
2. Include any of their evidence or reasoning for extreme scores or any that change the ranking.

For each dimension, team members should now have a scored backlog that looks like this:

Debt	Misalignment
Design System	5
FixMVC	5
Unit Tests	5
I18n	3
Zero Warnings	3
Dependencies	2
Folder Mess	2
Python3	1
Markdown	0
Permissions	0
Sockets	0
Sync Reliability	0
Window Memory	0

Once this is completed for each dimension, combine them into the larger Dimension Scoring spreadsheet.

	Forces that indicate we should pay						Forces that indicate we should stay with debt					
Debt	Visibility	Benefit of Paying	Misalignment	Volatility	Cost of Staying	# Resistance	Size	Cost of Paying	Difficulty	Uncertainty	Benefit of Staying	Regression
Dependencies	0	1.00	2	1	1.00	1	2	2.00	2	0	1.00	2
Design System	3	4.00	5	3	3.50	4	1	1.00	1	1	0.50	1
FixMVC	0	2.50	5	3	4.00	5	3	3.00	3	1	1.50	2
Folder Mess	0	1.00	2	5	4.50	4	1	1.00	1	1	1.00	1
I18n	5	4.00	3	5	2.50	0	5	4.00	3	0	0.50	1
Markdown	5	2.50	0	1	1.00	1	1	1.00	1	1	1.00	1
Permissions	3	1.50	0	1	2.50	4	2	3.00	4	3	3.50	4
Python3	0	0.50	1	4	3.50	3	3	2.00	1	0	1.00	2
Sockets	4	2.00	0	1	3.00	5	5	5.00	5	4	4.50	5
Sync Reliability	2	1.00	0	1	1.00	1	2	2.50	3	5	5.00	5
Unit Tests	0	2.50	5	5	5.00	5	5	3.00	1	0	0.00	0
Window Memory	5	2.50	0	2	1.00	0	3	2.00	1	1	1.00	1
Zero Warnings	0	1.50	3	4	2.50	1	3	2.00	1	0	0.50	1

This should be completed before the Kickoff Follow-up Meeting begins.

Meeting Agenda

Total time: 1 hour

Part 1: Review the priority (15-20 minutes)

On the "Prioritize" tab of the spreadsheet, see the net forces that resulted from the scoring.

Debt	Tɪ Pay or Stay	# Net Force	# Joy	Net Force with Joy	Bubble Size
Design System	Pay	3.55	2.00	3.01	2
Unit Tests	Pay	2.93	2.00	2.39	2
Folder Mess	Pay	2.32	5.00	3.14	5
FixMVC	Pay	1.66	5.00	2.18	5
I18n	Pay	1.66	3.00	1.57	3
Zero Warnings	Pay	0.72	1.00	0.37	1
Markdown	Pay	0.66	4.00	1.78	4
Python3	Pay	0.50	3.00	0.95	3
Window Memory	Pay	0.24	1.00	0.00	1
Dependencies	Stay	-0.42	5.00	0.95	5
Permissions	Stay	-1.98	2.00	-1.98	2
Sockets	Stay	-2.33	5.00	-1.13	5
Sync Reliability	Stay	-3.35	4.00	-2.43	4

Also note the bubble chart that puts high priority items on the upper right and lower ones on the bottom left.

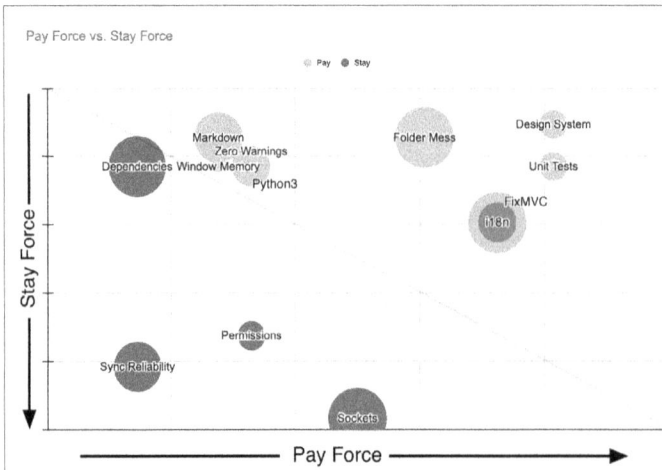

1. Review the Prioritize sheet and the bubble chart that shows each tech debt's Pay Forces and Stay Forces (and the net force).
2. If your team has a gut feeling that a location of a tech debt item in the chart is wrong, go back and review its underlying scores. Adjust the scores if there is consensus to do so.
3. Use the Joy column to express subjective feelings about the debt.
4. Prioritize discussing debt items with the most support for being paid first. Use the two net forces columns to help decide. This is a priority for discussion—you may choose to do work on debt further down the list that will help move it up for next time.

Part 2: Make a plan for each item in priority order (30-40 minutes)

1. In the Tech Debt Planning Google Sheet, review each item. Look at the underlying scores. Visualize items with a radar chart to get a better sense of what kind of tech debt you are dealing with. See Chapter 25 for examples.
2. Make a plan for the next quarter that takes your scores into account. Refer to Chapter 26 ("Decide What to do About Your Tech Debt") for guidance on this process and Appendix 6 ("Tech Debt Dimension-Driven Strategy Catalog") for ideas on how to align your plans with your debt scores.
3. Assign tasks that should be completed before the next recurring meeting.
4. If any of the items belong in your main bug/feature tracking system's backlog, move them there.

Publish the plan in a location where your team can easily refer to it. Your next meeting (three months from now) should start with a review of it.

Tech Debt Recurring Meeting Sample Agenda

Note that this meeting is discussed at the end of Chapter 26, "Decide What To Do About Your Tech Debt." It starts with a review of the last meeting, but then proceeds like the kickoff follow-up.

Purpose

To celebrate wins, review our progress in reducing tech debt, update the tech debt backlog, and make a plan for the next quarter.

Asynchronous Pre-work

1. At the end of the previous meeting, some team members are assigned to work on the plan.
2. Before this meeting, that work should be done using whatever processes you use. The assigned team members should be prepared to give an update.
3. All team members should have added new tech debt items they want to discuss at this meeting to the backlog.

Meeting Agenda

Total time: 1 hour

Part 1: Review the tasks assigned at the previous meeting (5-10 minutes)

1. For each task, the team member assigned to it gives a status update.
2. Celebrate any tech debt that has been cleared.
3. Plan to keep track of paid debt for a few quarters to collect evidence that longer-term benefits are being delivered.

Part 2: Update the backlog based on new evidence or other changes (15-20 minutes)

1. Quickly review each existing item in your backlog and update its dimension scores if necessary. It's likely that the work your team has done may have affected its score; if so, it should be recalculated. For example, if you added new requirements to your tech debt guides, some items may now be more misaligned (i.e., have a higher Misalignment score). If we did a spike and confirmed that our idea to resolve a debt will work, then the item's uncertainly will be lower.
2. Review any new backlog items and score them.

Part 3: Make a plan for each item in priority order (20-30 minutes)
This is the same thing we did in the Kickoff Follow-up Meeting, but here you should take into account any item scores that were revised and new backlog items.

1. In the Tech Debt Planning Google Sheet, review each item. Look at the underlying scores. Visualize items with a radar chart to get a sense of what kind of tech debt you are dealing with. See Chapter 26 for examples.

2. Make a plan for the next quarter that takes your debts' (revised) scores into account. Refer to Chapter 26 ("Decide What to do About Your Tech Debt") for guidance on this process and Appendix 6 ("Tech Debt Dimension-Driven Strategy Catalog") for ideas on how to align your plans with your debt scores.

3. Assign tasks that should be completed before the next recurring meeting.

4. If any of the items belong in your main bug/feature tracking system's backlog, move them there.

Publish the plan in a location where the team can easily refer to it. Your next meeting (three months from now) should start with a review of it.

APPENDIX 5

Tech Debt Dimension Scoring Guides

I. Visibility

If this debt was paid, how visible would it be outside of engineering?

(Note that in the Guidance column, "the product" refers to the deployed product itself, not the code for it.)

Score	Guidance
0	The product would not be changed in a way we (engineering) could see.
1	The product would have benefits to engineering.
2	The product would have benefits to internal teams.
3	Customers would get benefits, but not high-priority ones.
4	The product would benefit internal teams in a way that directly helps them earn revenue.
5	Customers would receive benefits that are identified on the near-term roadmap.

Visibility is a measure of how likely it would be for our customers and stakeholders to understand and value the debt being paid. When tech debt is very visible, it indicates there's a higher benefit to paying the debt. For example, if rewriting a system to use a new language feature would also fix a dozen reported bugs that we can't fix otherwise, then that debt has high visibility.

I18n preparation projects (to externalize strings from the code) are often highly visible because they make it possible to localize the application into foreign languages. The visibility can be high if clearing the debt causes the customer benefits or if it can be closely coupled with work that does.

Higher scores should be backed by evidence. To achieve a score of 5, this debt, when resolved, should help to implement stories on the current sprint, high-priority bugs on the backlog, or current roadmap items that are scheduled for the next six months.

To receive a 4, there should be evidence in closed-lost opportunities for your sales team. Low scores indicate that the product's new behavior, if any, mostly benefits us or our colleagues, not customers.

2. Misalignment

If this debt was paid, how much more would our code match our engineering values?

Score	Guidance
0	The code would not be more aligned with our values.
1	The code would be more aligned with values that we hold but are not formally documented.
2	The code would more closely match widely accepted engineering principles, but we haven't decided whether to follow them.
3	The code would more closely match engineering requirements that are documented in our coding guides.
4	The code would more closely match high-priority engineering requirements that are documented in our coding guides.
5	The code would more closely match engineering requirements that are company-wide and mandated from the C-level.

This score represents the extent to which our code matches our documented values. When a code is very misaligned, it indicates there is a higher benefit to paying the debt.

Higher scores should be backed up. Add a link to the documented requirement to your backlog item if it has a high misalignment score.

3. Size

If we knew exactly what to do and there were no coding unknowns at all, how long would the tech debt fix take?

Score	Guidance
0	An hour
1	A day
2	A few days
3	A week
4	A few weeks
5	Months

This number should represent the best-case scenario for your plan to fix this debt. This is like a T-Shirt size or a Story Point estimate, and we should provide the same kind of evidence here that we do for our stories. Even low scores should be justified, because it's very likely we'll decide to pay a debt if it takes a trivial amount of time.

The total cost to paying debt is this value combined with its difficulty (next section). If the scores for these two dimensions are high, then that would be a force pushing us back from paying it.

4. Difficulty

What is the risk that work on the debt takes longer than represented in the Size score because we won't know how to do it?

Score	Guidance
0	No risks; we know exactly how to fix the debt, and our Size estimate is very likely to be correct.
1	There are small risks.
2	Could take 25 percent longer
3	Could take 50 percent longer
4	Could take 75 percent longer
5	Could take twice as long or more

This score represents the risk that you might not actually know how to clear the debt once you get started on it. Higher numbers should be justified with a list of unknowns. Smaller numbers should be justified by a plan that team members agree has minimal or no risks.

5. Volatility

How likely is the code to need changes in the near future because of new planned features or high-priority bugs?

Score	Guidance
0	We will never change this code. We might even delete it.
1	No current plans to change this code.
2	Low priority bugs are in this area of the code.
3	Bugs are likely in this area of the code that we will need to fix.
4	Current plans are on the roadmap to change this code.
5	Current plans are on the roadmap to change this code many times.

Volatility represents how frequently or likely the code with tech debt will slow us down in the near future. It is related to Visibility but is meant to measure the consequences of not fixing debt. Visibility is the benefit we can attach to debt if it resolves items on our roadmap to address it.

Higher scores should be backed by evidence that this code has high-priority bugs or roadmap items that will cause you to encounter its debt. You can use lists of recent changes in your repository logs as evidence that this code needs constant maintenance.

Also note whether your roadmap plan is to completely reimplement how a feature works (for business reasons). This wouldn't result in a score of 5, because the product plan might be to delete it. If that's the case, you would not want to pay any debt in it, so rate it 0.

For example, you might have a complex screen written in Angular, but now you use React and you want to convert it. But if the PM is actually redesigning the screen to be completely different for product reasons, you would just want to wait and write the new screen in React and then delete the old one.

6. Resistance

How hard is it to change this code if we don't pay the debt?

Score	Guidance
0	This code is easy to maintain and change in the ways we plan to update it.
1	The overall architecture is fine, but the code is somewhat risky to change without causing more problems.
2	The code is very hard to change without causing problems.
3	The architecture is showing signs of stress. It's acceptable but trending worse.

| 4 | The architecture won't support us long term. |
| 5 | This code can't do what we need it to do without being rewritten. |

This score represents the cost a debt creates if you try to work with it. As always, higher scores should be justified with evidence, such as proof that the code has high complexity and low test coverage.

Resistance is related to Volatility, which measures the likelihood that we will pay a debt's cost; Resistance is the *size* of that cost. Taken together they determine the interest payments on debt.

Generally, when I refer to code being "resistant to change," I mean individual lines of code. Things like long, complex, undertested functions or unorganized code are more code-oriented problems. When I refer to unsupportive architecture, I'm talking about problems that can't be fixed by having cleaner code—a bigger fix is required. For instance, if you need to change your database from SQL to No SQL, the current system will resist your updates until you do that. Your workarounds in SQL may be fine, but they are the interest payment you are making until you pay the debt.

7. Regression

How bad would it be if we introduced new bugs in this code when we try to fix its tech debt?

Score	Guidance
0	We don't care if we cause regressions in this area of code.
1	There may be some minor regressions but customers don't rely on the current behavior of this code.

2	The behavior is nearly correct, and some customers rely on it.
3	The behavior is currently correct, and some customers rely on it.
4	This is an important area that is core to our product.
5	The behavior is currently correct, and customers rely on it widely. Or important customers do.

This score represents how essential the product features the code supports are. Regression is different from Visibility. Visibility is about the future and represents how much a change would be appreciated by non-engineering stakeholders (particularly customers). Regression represents how essential the code we're changing already is. It represents a benefit to not paying debt because whenever we change something that's working, there's a chance we could break it.

To get a high score, there should be evidence of use via analytics or references to the functionality in marketing materials and sales pitches.

8. Uncertainty

How sure are we that our tech debt fix will deliver the developer productivity benefits we expect?

Score	Guidance
0	We are sure that we will immediately boost developer productivity.
1	We are sure that we will boost developer productivity within six months.
2	We are sure that we will boost developer productivity eventually.
3	There are known, small risks to our plan to get developer productivity benefits from the fix.

| 4 | There are known and potentially large risks to our plan to get developer productivity benefits from the fix. |
| 5 | We don't know if the plan will increase developer productivity. We suspect there are unknown unknowns. |

This score represents our level of certainty that a tech debt fix will increase developer productivity even if it is otherwise successful. A high score for uncertainty indicates that we should not pay this debt.

A small pilot project with measured results (such as a spike) is a good, low-risk way to lower uncertainty.

APPENDIX 6

Tech Debt Dimension-Driven Strategy Catalog

This appendix expands on the ideas in Chapter 26 ("Decide What to do About Your Tech Debt"). In that chapter, I argued that you should ensure that your plan to fix a tech debt aligns with the debt's dimension scores. In other words, you should have strong evidence for any decision you make about debt, whether you decide to fix it or to leave it alone.

(One related note before we continue: I assume by now that you are familiar with the eight dimensions of tech debt: **visibility**, **misalignment**, **size**, **difficulty**, **volatility**, **resistance**, **regressions**, and **uncertainty**. If you aren't, the content below will probably not be very useful, since the strategies I cover are each based on the results of your dimension scores. So if you haven't yet read Part 3, go back and do so before continuing.)

To refresh, here's the diagram showing how the eight tech debt dimensions act as a force to pay a debt or leave it alone:

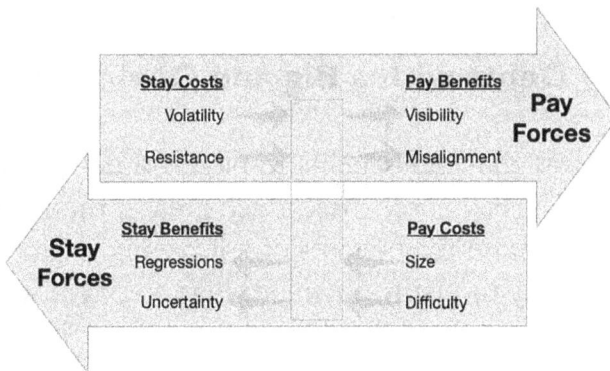

For example, imagine you had code that was very hard to change (high resistance) and was also in an older programming language you wanted to replace (high misalignment). These two facts strongly suggest you should pay the debt, i.e., serve as strong Pay Forces. If you were sure you knew how to do it (low difficulty) and that the resulting code would definitely bring high developer productivity (low uncertainty), that would also indicate a benefit to paying, since these two Stay Forces scores are low.

In this appendix I have tried to go beyond simply positing scenarios where it makes sense to pay or not. I also recommend specific actions to take if you do decide to pay, as well as actions you should take with debt that you probably shouldn't pay but don't want to totally ignore. The general strategy is to look at "stay" score that is high or a "pay" score that is low and try to do small projects to change those numbers. Over time this would make that item have obvious benefits and lower costs. At some point, we'd just pay it off.

One of my assumptions here is that you have set aside some portion of your budget to address tech debt issues, as I described in Chapter 20 ("Make and Spend a Budget to Pay Back Tech Debt"), and that your main decision now is how to allocate it. Accordingly, my references to "your budget" below refer to the budget you have set aside for engineering-led activities, like tech debt.

Pay Debts with a Big and Obvious ROI

When it's clear that the benefits to paying a debt and the cost of not paying it (Pay Forces) are much greater than the costs of paying the debt and the benefits of not paying it (Stay Forces), it should be easy to justify payment. Plan to pay it within your budget.

This sort of debt should produce highly visible benefits to customers and obvious increases in developer productivity. It might also be small, trivial to do, and easy to verify. It's the easiest kind of debt to sign off on.

Ignore Debts with Low (or Negative) ROI and Big Risks

Similarly, if it's clear from your scoring that a debt is not having much of an effect on developer productivity or user experience and it would cost a lot to address it, ignore it. Keep it in your tech debt backlog but in a low position, as a reminder of why you aren't doing anything about it.

Work within Product-led (Roadmap) Initiatives

If a debt's **Visibility** dimension is very high, it's likely that your PM has features planned that would be easier to implement if you paid the debt. If that's the case, you should develop two different plans for implementing the features. The first should address the debt, and the other should be a workaround that enables the features (though without paying the debt).

If the plans have similar costs and risks, hopefully you can argue in favor of the one that pays the debt. But if the debt-paying plan will take longer or has some risks, engineers should be willing to contribute part of their engineering-led budget to what would otherwise be a product-led project (and would use part of the PM budget).

The best way to do this is to pay the debt before feature work starts, as part of preparation, but not at the same time. This process works best for debt items that are in the near-term roadmap (often referred to as "Next"), not the imminent one ("Now").

The engineers assigned to pay the debt could then transition to the feature when they are done. Their work on the debt would count against the engineering-led budget, and their work on the feature would count against the product-led budget. The PM should be indifferent since their cost is the same. If the debt project fails, the team can fall back to the plan where the feature is implemented without paying the debt.

Build Debt Observability to Show Visibility

If a debt's **Visibility** score is low, but you feel like it is underestimated, then you should work to produce more evidence. Build in more observability that makes the effect of the debt on your customers clearer.

(See Chapter 9, "Make the Effects of Tech Debt More Visible," for more on this topic.) If you are right, you can adjust the score in the next meeting and see if it changes the situation.

Find and Tag Issues Related to Debt

If a debt's **Visibility** dimension is scored low but you have seen bugs or support cases that are being caused by the debt, add a unique tag to your bug-tracking system and tag the items in a way that makes this clear. Then search for older items and tag them too, even if they've been closed and marked as fixed. Use the count and severity of bugs with these tags as evidence that **Visibility** should be scored higher.

Spike to Reduce Risk

If a debt's **Difficulty** score is very high, create a timeboxed spike to explore the problem and, as a result, get a more accurate estimate of its **Difficulty**. This might change the debt **Size** (in either direction), but either way it will reduce the chance that the **Size** estimate is wrong, thus reducing the **Difficulty** score.

Similarly, if you scored **Uncertainty** very high because you don't know if paying the debt will increase productivity and reduce build times, do a time-boxed spike to test out your plan to fix it. Experimenting in a small area of the codebase should give you an answer fairly quickly.

Create a New Coding Standard

If an item could never have high **Visibility** and is driven more by engineering concerns, but there is no agreed-upon team standard to measure it against, then you should be scoring it low on **Misalignment** as well. But you could increase that score by deciding on and documenting a new coding standard. Having this in place will also enable you to use one of the Misalignment-driven strategies in this catalog.

Those strategies work especially well for debt items you are happy to tackle via small, regular fixes over time, since your changes will likely be gradual and less disruptive. Once your new standards are documented, they should be enforced in code reviews. Your rule should explain when it's appropriate to follow them and when it's okay to ignore them. For example, you might only convert a file from an old programming language to a new one when you have to make major changes to the file anyway.

If you are following this, then update the item's **Size** score in each meeting to reflect that the debt is shrinking. At some point this might tip the ROI enough to just finish it in a project.

Automate a Coding Standard

If **Misalignment** of a tech debt item is very high and not improving, automate the standard so that it is more likely to be addressed in PRs. You can automate inside the IDE with a linter or automated formatter. You can also do it with a commit rule when a PR is created or as a pre-merge check in your CI system.

Build Debt Observability to Show Misalignment

Another strategy to consider if **Misalignment** is high but not improving: creating a dashboard that makes this obvious to the whole team. This works well for displaying codebase metrics such as the combination of coverage and complexity, described in Chapter 9 ("Make the Effects of Tech Debt More Visible"). It also works well during large-scale language migrations, like going from Objective-C to Swift or CoffeeScript to TypeScript.

Add Tests

If **Resistance** is scored high because an area is untested, add tests to make the area less risky to change. This is especially important when the cost of **Regressions** is high. It is important to add tests to code that also has a high chance of needing changes (i.e., has a high **Volatility**). The existence of code that is risky to change ultimately doesn't matter if you aren't going to change it.

Make Code More Compliant

If **Misalignment** is high because you already have a coding standard but your current code doesn't follow it, allocate some time to address an area of it. This is especially important when roadmap items in the near term ("Next" quarter) will be sped up if you fix these problems now.

Fix Small Issues Immediately

If you run into small things as you work on a roadmap feature or bug, just fix them in the same PR and don't add them to the backlog. Make small mechanical refactorings to make an area less complex or add useful comments. Change existing code in uncontroversial ways. This works well for large **Misalignment** problems that you can't justify fixing all at once. Couple this with a dashboard to show that progress is being made.

The goal is to decrease **Size, Difficulty** (the risk that your estimate is wrong), and **Uncertainty** (the risk that the fix doesn't make the code easier to change) over time. At some point the project will be small and easy enough to justify a small project to finish it.

Propose Removing a Feature

If you have evidence that **Regression** risk is very low because users don't actually use a feature that has a lot of debt, propose to the PM that it be removed. Long-lived projects often have a lot of these kinds of things that aren't used but cause work to keep up to date.

Create an Onboarding Project

If you have some debt that a new hire could work on to learn your codebase, write a project specification for them that involves the debt. I recommend having them work on one that's driven by **Misalignment** and **Resistance** and that has a low chance of **Regression**. It would also be best if the project is small or could be done incrementally. The goal is to train the new hire, so having them bring your code into compliance with your standards and make more frequent PRs is a great way to do this.

Automate a Step of a Manual Process

This works especially well when a process is done often and also gets out of date or is hard to understand. Instead of formally documenting the step, see if you can automate it incrementally.

Replace Tech Debt with a Third-Party Dependency

When a system is old, there is often something in it that was novel when the project was started but can now be implemented better in a well-documented and widely used open-source project. If your code has high **Volatility** and **Resistance**, which are the ongoing costs of not fixing tech debt, consider replacing it with a library. This, of course, has its own risks, so use the principles of substitutability and consider having parallel implementations during a migration period.

This works best in projects that are looking to increase business value by reducing ongoing costs, driven by operational excellence. Outsourcing (to a library) is a common way to achieve that.

All Hands on Deck to Clear Tech Debt

If you have a very large and obvious tech-debt problem, negotiate with your PM to spend a short period just dealing with tech debt issues. (See Chapter 9, "Make the Effects of Tech Debt More Visible," and Chapter 14, "Remove Tech Debt When It's

Your Decision," for examples of how to do this.) Your pitch is more likely to be successful when your PM and Design don't have anything pressing to work on. In return, you will realign the budget allocation by spending more time on product-led work in the future.

Gather Support to Do Something Big

Sometimes the ROI for paying a debt is clear and the costs of not paying are high, yet the costs of paying are *also* high and there isn't a good way to make incremental progress. In this scenario, your only option might be to elevate things to leadership to get more support. This is a last resort, though. You should first try things like spikes and making incremental progress, even if they only help you learn more about the extent of the problem. See Chapter 30 ("Give Big Rewrites Enough Support [or Don't Do Them])" for additional information on this.

Acknowledgements

I have been lucky to be taught by, go to school with, and work with software developers who have a serious and thoughtful view on the craft of programming and have taken the time to teach me so much of what is in this book.

No list would be exhaustive, but I want to thank some who came to mind as I wrote this book (in the order that I met them). Thanks to my teachers: Mr. Penner, Chris Lent, Jeff Hakner, Dr. Simon Ben-Avi, and Rich Hickey. To my classmates: Sari Eskildsen and Robb Horn, and my colleagues Frank Rose, Philip Brittan, Kira Sirote, Mike Tovino, Matt Bruce, Jennifer Rippel, John Ashworth, Mat Baskin, Johanna Voolich Wright, Aurangzeb Agha, Bill Bither, Steve Hawley, Rick Minerich, Sage Mitchell, Don Frehulfer, Dan Lew, Emily Chapman, Stephanie Marshall, and Ron Sharon-Zipser. I might not have finished if my friend, Adelia Thomas, hadn't said just the right thing at the right time to stop me from giving up. I would also like to thank my mom, Josephine, who got me my first computer, and her twin sister, my Aunt Grace, who helped me get my first job programming.

In writing this book, I have been part of the Useful Books author community started by Rob Fitzpatrick and attended writing accountability sessions led by Brian Hall and Adam Rosen. My thanks to them and my fellow community members, whose books I can't wait to read.

Two community members, Adam Rosen and Marjorie Turner Hollman are also professional editors who I hired to help me. Adam served as my developmental editor and was instrumental in shaping the structure of the book. Marjorie and her staff

(Pat Nickinson, PhD and Jon Hollman) did the final editing and proofreading. I also relied on Alex Liebowitz's keen eye and technical knowledge on the final version. If you find any errors, though, it's probably my fault because I couldn't stop futzing with the book.

The community at Blogging for Developers, led by Monica Lent, was invaluable in getting me to blog more, which is how I developed a lot of the content in this book. It was also there that I met Gergely Orosz, who invited me to contribute an excerpt to his *Pragmatic Engineer* newsletter. His readers' comments helped shaped the book.

And of course, I stand on the shoulders of giants like Ward Cunningham (who coined the term "tech debt"), Kent Beck (whose book taught me about unit testing), Martin Fowler (whose book taught me about refactoring), and Michael Feathers (whose book taught me how to work with legacy code). I would also like to thank Dr. Barbara Liskov, of the Liskov Substitution Principle, for influencing nearly every line of code that I write—but especially the code I write to pay tech debt without breaking everything.

www.ingramcontent.com/pod-product-compliance
Lightning Source LLC
Chambersburg PA
CBHW061157240326
R18026500001B/R180265PG41519CBX00024B/41